STILL THE IRON AGE

STILL THE IRON AGE

Iron and Steel in the Modern World

VACLAV SMIL

Amsterdam • Boston • Heidelberg • London • New York • Oxford
Paris • San Diego • San Francisco • Singapore • Sydney • Tokyo
Butterworth-Heinemann is an imprint of Elsevier

Butterworth-Heinemann is an imprint of Elsevier
The Boulevard, Langford Lane, Kidlington, Oxford OX5 1GB, UK
50 Hampshire Street, 5th Floor, Cambridge, MA 02139, USA

Notices
Knowledge and best practice in this field are constantly changing. As new research and experience broaden our understanding, changes in research methods, professional practices, or medical treatment may become necessary.

Practitioners and researchers must always rely on their own experience and knowledge in evaluating and using any information, methods, compounds, or experiments described herein. In using such information or methods they should be mindful of their own safety and the safety of others, including parties for whom they have a professional responsibility.

To the fullest extent of the law, neither the Publisher nor the authors, contributors, or editors, assume any liability for any injury and/or damage to persons or property as a matter of products liability, negligence or otherwise, or from any use or operation of any methods, products, instructions, or ideas contained in the material herein.

ISBN: 978-0-12-804233-5

British Library Cataloguing-in-Publication Data
A catalogue record for this book is available from the British Library.

Library of Congress Cataloging-in-Publication Data
A catalog record for this book is available from the Library of Congress.

For Information on all Butterworth-Heinemann publications
visit our website at http://store.elsevier.com/

CONTENTS

PREFACE AND ACKNOWLEDGMENTS

My books are expressions of my preference for writing about fundamental realities, be they natural or anthropogenic, and their complex interactions. That is why I have written extensively on the Earth's biosphere and its transformations by humans, on production of foods and changing diets, on energy resources and on material foundations of our civilization. Besides dealing with these matters in systematic, universal and generalized manner (the best example would be *General Energetics*, *Global Ecology*, *Feeding the World*, *Energy in Nature and Society*, *Harvesting the Biosphere* and *Making the Modern World*) I have taken some closer looks, writing books focusing on specific fundamentals of modern civilization: on wood and other biofuels (*Biomass Energies*), oil (*Oil: A Beginner's Guide*), natural gas (*Natural Gas: Fuel for the 21st Century*), ammonia (*Enriching the Earth*), Diesel engines and gas turbines (*Prime Movers of Globalization*), and meat (*Should We Eat Meat?*). This book is simply a continuation of my efforts to deal with such fundamental realities and it has been on my list of to do items since the early 1990s when I began to study the long history and remarkable accomplishments of iron smelting and steel making.

Those who appreciate the physical foundations of modern societies do not need any convincing about the topic's importance. Those who think that mobile phones and Facebook and Twitter accounts are the fundamentals as well as the pinnacles of modern civilization might find the book about iron and steel inexplicably antiquated: their realities appear to be purely silicon-based. But that, of course, demonstrates deep lack of understanding of how the world works. Modern civilization could exist quite well without mobile phones and "social media"; indeed, in the first instance it did so until the 1990s (beginning of large-scale adoption of cellphones) and in the second instance until the late 2000s (when the Facebook membership took off). In contrast, none of its great accomplishments—its surfeit of energy, its abundance of food, its high quality of life, its unprecedented longevity and mobility and, indeed, its electronic infatuations—would be possible without massive smelting of iron and production (and increasingly also recycling) of steel.

In the 1830s Danish archeologist Christian Jürgensen Thomsen (1788–1865) distinguished three great civilization eras based on their dominant hard materials, with the Bronze Age following the Stone Age and preceding the Iron Age (Thomsen, 1836). Transition from stone to bronze began about 3300 BCE in the Near East and just a bit later in Europe, the onset of Iron Age was around 1200 BCE but it took another 700–1000 years before the metal became dominant throughout Asia and Europe. When Thomsen made his division, the Iron Age was mostly 2000–2500 years old, but the time of the greatest dependence on the metal was still to come, and at the beginning of the twenty-first century no other material has emerged to end that dominance. Ours is—still and more than ever—the Iron Age although most of the metal is now deployed as many varieties of steel, alloys of iron and carbon (typically less than 2% C) and often of other metals that impart many desirable qualities absent in pure elemental iron.

The great nineteenth-century surge in iron smelting and steel production continued during the twentieth century as the long-lasting US technical leadership shifted to Japan after 1960. Four decades later the rapid expansion of China's economy brought the iron and steel output to unprecedented levels during the first decade of the twenty-first century. By 2015 iron ore extraction was more than 2 billion tonnes (Gt), the mass surpassed only by the annual output of fossil fuels and bulk construction materials; pig (cast) iron production (smelting of iron ores in blast furnaces) rose to more than 1 Gt; and the global steel output (from pig iron or from recycled metal) reached about 1.7 Gt. That output was about 60 times higher than in 1900, and roughly 20 times larger than the aggregate smelting of aluminum, copper, zinc, lead and tin. And in per capita terms worldwide steel output rose by an order of magnitude, from 20 kg/year in 1900 to about 230 kg/year by 2010.

Perhaps the best way to stress the importance of steel in modern society is to note that so many components, parts, machines and assemblies are made of steel and that just about everything around us is made or moved with it. Although naked steel is not uncommon—ranging from such small items as needles, pins, nails, construction, laboratory and medical tools to slender broadcasting towers, wires, cables, rails and bridge spans and girders—most of the metal incorporated in modern products is hidden (inside structures as reinforcing bars in concrete, skeletons of large buildings, inside machines as engines and turbines or underground in piles, pipelines, tunnels and mine props) or covered by layers of paint (welded ship hulls, construction machinery, cars, appliances, storage tanks).

The list of items and services whose reliability and affordability have been made possible by steel is nearly endless as critical components of virtually all mining, transportation and manufacturing machines and processes are made of the metal and hence a myriad of non-steel products ranging from ammonia (synthesized in large steel columns) to wooden furniture (cut by steel saws), and from plastic products (formed in steel molds) to textiles (woven on steel machines). And steel's qualities have improved as its uses have spread. High strength steel forms the skeletons of skyscrapers and prevents ceilings from caving in in deep coal mines; when alloyed with Cr, Mn, Ni, V, and other metals specialty steels can be used in corrosive environment or operate under high temperatures and pressures, such as stainless steel blades of large steam turbines that are the world's dominant generators of electricity. Woven steel cables suspend the world's longest bridges and steel makes up the bulk of products ranging from oceangoing vessels, cars, refrigerators, and productive assemblies ranging from large refineries to massive offshore oil drilling rigs.

The verdict is easy: although the last two generations have seen an enormous amount of attention paid to many admirable advances in electronics, affluent civilization without microchips and ubiquitous telecommunication is perfectly possible; in contrast, modern high-income, high-energy societies would be impossible without steel. First things first: ours is still very much the Iron Age, and this book will trace its genesis, slow pre-industrial progress, revolutionary advances during the nineteenth century and their further magnification during the past five generations. Afterwards I will look at the patterns of modern steel production, the metal's ubiquitous uses, potential substitutions, advances in relative dematerialization and, without any counterproductive time-specific forecasting, I will close the book with a brief appraisal of steel's possible futures.

My thanks to six people who have taken care of this book's illustrations: Anu Horsman and Ian Saunders arranged the rights for Corbis images; Hiroyuki Tezuka and Naoyuki Haraoka secured all Japanese photographs; and Evan Kuz and Neil Smalley took care of the graphs. I also thank to several anonymous reviewers whose comments helped to shape the book's final content.

PREVIOUS WORKS BY THE AUTHOR

China's Energy
Energy in the Developing World (edited with W. Knowland)
Energy Analysis in Agriculture (with P. Nachman and T.V. Long II)
Biomass Energies
The Bad Earth
Carbon Nitrogen Sulfur
Energy Food Environment
Energy in China's Modernization
General Energetics
China's Environmental Crisis
Global Ecology
Energy in World History
Cycles of Life
Energies
Feeding the World
Enriching the Earth
The Earth's Biosphere
Energy at the Crossroads
China's Past, China's Future
Creating the 20th Century
Transforming the 20th Century
Energy: A Beginner's Guide
Oil: A Beginner's Guide
Energy in Nature and Society
Global Catastrophes and Trends
Why America Is Not a New Rome
Energy Transitions
Energy Myths and Realities
Prime Movers of Globalization
Japan's Dietary Transition and Its Impacts (with K. Kobayashi)
Harvesting the Biosphere
Should We Eat Meat?
Power Densities
Natural Gas

CHAPTER 1

Iron and Steel Before the Eighteenth Century

Slow Adoption, Artisanal Production, and Scaling-Up

We can only guess at the beginnings of metal smelting in Neolithic societies. Were the minerals containing metals with low melting points accidentally present in or near fire pits used for heat or for searing meat, and did their melting attract the attention of people tending the fires? Did curiosity lead people to throw colored minerals into fires to see what will happen? Or did the discoveries of naturally occurring nuggets, crystals, or lumps of native metals (copper, gold, silver, lead, tin) lead to deliberate experimentation with minerals (metallic ores) that contained small particles of those elements? And once melting of some materials was discovered, were not attempts at their deliberate smelting almost inevitable? Craddock (1995) thinks that was almost certainly the case.

What we know for certain is that the earliest evidence of exploiting native metal—as beads of malachite and native copper in southeastern Turkey—goes as far back as 7250 BCE (Scott, 2002). Because copper often occurs with arsenic and the eutectic point (the lowest melting temperature) of Cu–As alloys is just 685°C, the first bronzes, encountered at the end of the 4th and the beginning of the 3rd millennium BCE in many settlements in Mesopotamia, were variants of this natural combination. Bronzes that eventually gave the name to the first metallic age were alloys of copper and tin (with the eutectic point at 910°C), and they were introduced throughout the region around 3500 BCE (De Ryck, Adriaens, & Adams, 2005).

While the protometal cultures (up to the 5th millennium BCE) were confined to Mesopotamia and southeast Turkey, the copper age (5th millennium BCE) extended into the Nile Valley, Southeastern Europe, and the steppes north of the Black Sea, cultures of the early bronze age (4th millennium BCE) occupied large parts of southern and eastern Europe and reached eastward to the Indus Valley, the middle bronze age

Still the Iron Age.
DOI: http://dx.doi.org/10.1016/B978-0-12-804233-5.00001-4

(3rd millennium BCE) included nearly all Europe except for Scandinavia as well as parts of China, and the societies of the late bronze age (2nd millennium BCE) were found across Eurasia, from Portugal to Korea and from southern Siberia to India and Ethiopia (Chernykh, 2014).

But despite this lengthy smelting experience, the transition from the reliance on bronze to societies whose dominant metal was iron took typically many hundreds of years, in some cases an entire millennium. Again, the Middle Eastern civilizations pioneered the process, and again, the first manufactured iron objects were made from native iron: the oldest known iron artifacts are nine tubular Egyptian beads made from meteoritic iron (characterized by large crystal grain size and high nickel content), and reliably dated to about 3200 BCE (Rehren et al., 2013). The beads were discovered in 1911, and their shape was made by multiple cycles of rolling and annealing.

Perhaps the most surprising case of using meteoritic iron was discovered in 1818 when an English expedition, led by John Ross searching for the Northwest Passage, encountered the Inuit in northwest Greenland who had iron knives and iron spear points. The origin of these metal objects was puzzling, but eventually they were traced to iron flakes removed by hammer stones from meteorites of the Cape York fall (Wayman, 1988). Egyptian meteoritic beads predate the emergence of iron smelting by nearly 2000 years. Small iron objects are first documented from Mesopotamia of 2600–2500 BCE, but larger items and ceremonial weapons (the metal was still too rare to produce functional designs) became more common only after 1900 BCE; the metal was in everyday use after 1400 BCE and became widely affordable only after 1000 BCE.

BLOOMERY IRON

Smelting of iron followed the practices established for the production of color metals that had been going on in some parts of the Middle East for nearly 2000 years. Simple bowl-shaped hearths—shallow and usually clay- or stone-lined pits—were encircled by low circular clay walls. These walls were sometimes only knee high (Romans made most of their metal in furnaces no more than 1 m tall and less than half a meter of internal diameter), but in some parts of the Old World (including Central Africa) they eventually reached heights of more than 2 m (van Noten & Raymaekers, 1988). Furnaces were filled with charcoal and crushed (and often roasted) iron ore, and relatively high temperatures were achieved by blowing in air through tuyères, narrow clay tubes inserted near the surface level (see Appendix B for definitions of some major technical terms associated with the production of iron and steel).

Tuyères were connected to leather bellows to force air into the hearth and to raise smelting temperature. Small bellows were operated by hand, larger ones by a man's weight (using a treadle or a rocking bar), and the most powerful bellows were eventually powered by waterwheels. Temperature inside these charcoal-fueled furnaces usually did not reach more than 1100–1200°C (and often it was less than 900°C), high enough to reduce iron oxide and far from enough to melt the metal and produce liquid iron (pure Fe liquefies at 1535°C): the final product of this smelting was a bloom, a spongy mass made up of iron and iron-rich slag composed of nonmetallic impurities (Bayley, Dungworth, & Paynter, 2001). Hence the common name of these furnaces, bloomery, and of the product, bloomery iron.

Modern experiments demonstrated a relatively narrow range of conditions required for successful smelting (Tylecote, Austin, & Wraith, 1971). When the conditions inside the furnace are insufficiently reducing there is no metal produced, just iron-rich slag, but when they are too reducing slag becomes too viscous and cannot be easily separated from the metal. Intermediate conditions produce a good bloom; most of the slag comes from iron ore, about 30% originates from siliceous furnace lining, and less than 5% is fuel ash (Paynter, 2006). Blooms made in the smallest early furnaces weighed less than 1 kg, more typical medieval range was 5–15 kg, and the bloom mass increased to 30–50 kg (or even to more than 100 kg) only with the introduction of taller furnaces and waterwheel-powered bellows.

Bloomery iron contained typically between 0.3% C and 0.6% C, and in Europe it was the only ferrous material available in significant quantities during the antiquity and until the later medieval period. The iron produced by bloomeries was consolidated and shaped by subsequent smithing: repeated reheating and hammering of the bloom was required to produce a mass of wrought iron that contained just 0.04–0.08% C and that was ductile, malleable, and weldable. Wrought iron was used to make an increasing range of weapons and utilitarian and ornamental objects, ranging from arrowheads to bolts and axes (Ashkenazi, Golan, & Tal, 2013; Barrena, Gómez de Salazar, & Soria, 2008), and modern metallurgical examinations find small amount of slag trapped in these products.

Bloomeries supplied all of Europe's iron during the continent's first notable increase of demand for the metal that started in the eleventh century—with the introduction of iron mail, originally as small metal plaques, later as hand-forged and riveted knots—and expanded during the twelfth and thirteenth centuries. There was an increase in the production of hand weapons (ranging from knives to maces) and helmets, as well as agricultural and transportation tools and implements, with iron turned into plows, pitchforks, sickles, hoes, cart axles, hoops (for casks, wagons, and windmills), and horseshoes. The first documented use of powerful forge tilt hammers driven by waterwheels dates from 1135 in the famous Cistercian monastery of Clairvaux. More iron also went into construction as bolts, grills, bars, and clasps, and in the thirteenth century metal bands were used in Notre Dame de Paris. A century later the papal palace in Avignon consumed 12 t of the metal (Caron, 2013).

Bloomery smelting was practiced by virtually all Old World cultures, and thousands of these simple, temporary hearths (sometimes with parts of walls still intact) were excavated in regions ranging from both Sahelian and sub-Saharan Africa (Haaland & Shinnie, 1985) to nomadic societies on the steppes of Central Asia (Sasada & Chunag, 2014), and from coastal Sri Lanka (Juleff, 1996; 2009) to Scandinavia (Olsson, 2007; Svensson et al., 2009) and Korea, where the practice may have been transferred from what is now Russia's Pacific coast region rather than from China where cast iron was dominant (Park & Rehren, 2011).

Most of the evidence of the earliest Euroasian iron smelting has been known for a long time, with numerous remains of simpler and lower structures (often called Corsican forges) and sturdier and taller furnaces (called Catalan forges) found from the Atlantic to the Urals. In contrast, new excavations of ancient bloomeries and new carbon datings have been changing our views on the development of iron metallurgy in Africa (Holl, 2009; Zangato & Holl, 2010). These findings indicate early smelting activities in regions ranging from the Middle Senegal Valley in the west to the Nile Valley in the east, and from Niger's Eghazzer basin to the Great Lakes region of East Africa, with the many dates going to more than 2500 years before present and with inferred furnace temperatures of 1100–1450°C.

Persistence of this smelting technique is attested by the fact that the Spanish bloomeries at San Juan Capistrano (built during the 1790s) were the oldest ironworks in California, and operating bloomeries survived in parts of England into the eighteenth century; in parts of Spain and in southern France they were still present by the middle of the nineteenth century. Bloomery smelting was just the first step in obtaining useful metal: the ferrous sponge mixed with slag had to be processed by being repeatedly worked (wrought) by alternate heating and hammering (requiring as many as 30–50 cycles) in order to remove the interspersed impurities and to produce wrought iron that could be forged into weapons, horseshoes, colter tips, nails, and other small iron objects. For centuries all of this hot and hard labor was done everywhere manually, and only the adoption of larger waterwheels made it possible to build mechanized forges using heavier hammers. Even so, this traditional combination of bloomeries and forges had its obvious production limits.

Being small-scale batch operation—every heat was terminated in order to remove relatively small masses of the solid bloom—iron smelting in traditional low-rise bloomeries could never supply large-scale demand for the metal in an economic way, and labor-intensive (and also highly energy-intensive) forging added to the cost (further increased by substantial losses of iron during the forging process). Not surprisingly, with rising demand some European bloomeries, exemplified by medieval German and Austrian *Stucköfen*, became taller (*Technisches Museum* in Vienna has a fine model). These furnaces still produced small masses of metal (*Stuck*) whose removal required tearing the front wall of the structure, but because the smelting process lasted a bit longer and because waterwheel-driven bellows supplied more powerful blast and temperatures in lower parts of the furnace were higher, the resulting bloom was often a mixture of sponge iron and steel.

BLAST FURNACES

Japan's traditional *tatara* is the best known example of a furnace that could be both a bloomery producing a solid sponge iron and a blast furnace yielding liquid iron or steel (Hitachi Metals, 2014; Iida, 1980). *Tatara* furnaces were rather low and rectangular (height of just over 1 m, width of 1 m, and length of 3 m) and (since the late seventeenth century) the blast was delivered by cross-blowing bellows. Given the scarcity of Japanese iron ores, abundant iron sands were charged with charcoal and the furnaces were operated in two modes, as *zuku-oshi* (pig iron-pressing) and *kera-oshi* (steel pressing, which will be described in the following section). Pig iron production used *akome* iron sand (with high titanium content, as 5% or more of TiO_2) which was added after the charcoal charge; the furnace's tuyères were placed low and at a low incline so the blast could penetrate the entirety of the furnace's lower part; and, in order to achieve high temperatures needed to produce liquid iron, the heats lasted 4 days. The liquid metal was cast, or, after cooling, taken to a workshop for decarburization to produce *sage-gane* (low carbon steel) used to make a variety of everyday items.

Experiments with a replica of a 1.3-m tall *tatara* furnace built in Miyagi prefecture and charged with magnetite ore (rather than with iron sand) confirmed that it reaches very high smelting temperatures (in excess of 1500°C) during the last stage of its operation, and found increasing metal yield with higher metal content in the ore (Matsui, Terashima, & Takahashi, 2014). But even with the best available ore the yield was no higher than 62%, and with poorer ores more than half of iron present in charged iron sand could be lost in slag. Analysis of the produced metal showed carbon content of 0.15% and silicon content of 0.03%, with negligible amounts of Mn, P, S, Ti, and Cu, corresponding to low carbon steel.

But the first liquid iron was produced long before Japanese *tatara* became common. Unlike in Europe, where solid-state reduction of iron in bloomeries dominated the metal's production until the late medieval period, there is only sparse and inconclusive evidence of bloomery smelting in ancient China, but plentiful evidence of producing liquid iron for casting. Chinese metallurgists, after centuries of experience with bronze castings, became the pioneers of liquid iron production during the Spring and Autumn Period (770–473 BCE), and smelting of relatively large quantities of liquid iron in fairly large furnaces is well documented during the Han dynasty (207 BCE–220 CE).

Chinese smelting took advantage of phosphorus-rich iron ores (with a lower melting point) and good refractory clays. The tallest refractory clay structures were just over 5 m high and they were often strengthened on the outside by vine cables or heavy timbers, were charged with nearly 1 tonne of iron ore, and the process yielded two tappings of cast iron a day (Needham, 1964; Wagner, 1993). In larger furnaces, reduction of iron oxide began already during the early stage of the descent through the combustion zone, and once the right temperature is reached iron will rapidly absorb carbon (its share rises to 2% and to as much as 4.3%); its melting temperature is lowered as the eutectic point falls to just 1150°C. Another essential component of the early Chinese success was the use of double-acting bellows whose strong air blast helped to raise the smelting temperature. A later Chinese innovation was to use coal packed around tube-like crucibles charged with iron and larger bellows powered by waterwheels.

Chinese mastered the casting of liquid iron into interchangeable stacked clay molds (introduced during the Bronze Age for bronze casting) as well as into iron molds (introduced about 2000 years ago) in order to mass-produce a variety of farming and manufacturing tools as well as thin-walled cooking pots and pans. Ancient Chinese ironmasters had also found ways to overcome the inherently brittle nature of cast iron: they discovered that prolonged heating of the alloy at high temperature causes it to lose its brittleness, a process we now explain by conversion of cementite to graphite embedded within ferritic or pearlitic matrix.

Adz and hammer heads, hoe blades, and spade and plough tips were produced by liquid iron casting by the middle of the first millennium of the common era, and excavations of worksites from the Han dynasty (at the very beginning of the common era) show production of large quantities of coins, belt buckles, horse bits, vehicle parts, and smaller decorative items (Hua, 1983). Some fairly large iron statues were also cast during the Han dynasty, and even larger ones came later. They included eight iron oxen used to anchor the ends of cables supporting the Pujin bridge in Shanxi built around 720 during the Tang dynasty—the largest ones were 3.32-m long and weighed 70 t— and a 40–50 t lion in Cangzhou in Hebei hollow cast in 953 (Derui & Hanping, 2011). But this early deployment of liquid iron production and skilled casting soon ended in a technical cul-de-sac. As with so many other innovations in ancient China, there were hardly any improvements for centuries to come and modern blast furnaces eventually arose from designs used in medieval Europe.

European production of liquid (cast) iron took off only with the diffusion of blast furnaces. After controversies regarding the dating of the oldest found blast furnace in Sweden (Lapphyttan, in the mining region of Norberg), it is now accepted that the furnace operated for two centuries starting in the second half of the twelfth century, putting the country's first liquid iron production two centuries earlier than previously assumed (Rydén & Ågren, 1993). The other region with early blast furnaces was the lower Rhine Valley, starting during the latter part of the thirteenth century (to distinguish them from *Stucköfen*, their German name was *Flüssöfen*, liquid-producing furnaces, or *Hochöfen*, tall furnaces), and there are numerous reference to the casting of cannons from fourteenth-century Italy (Buchwald, 1998; Williams, 2009). In England, the first reliably documented blast furnace dates only to 1491.

In Europe and North America, cast iron from these furnaces has been widely known as pig iron (see Appendix B). In the late medieval and early modern Europe, it was often used directly as a substitute for bronze to cast large objects, above all guns, but for most other uses it had to be first converted into wrought iron or steel (see the following section). As with the early Chinese structures, gradual increase of European blast furnace stack heights allowed for a longer contact between the ore and the fuel, raised smelting temperatures, and reduced the melting point and charcoal consumption (Williams, 2009).

Georgius Agricola (1494–1555) gives a description of the prevailing practice of the sixteenth-century iron ore smelting in his famous *De re metallica*:

> *A certain quantity of iron ore is given to the master, out of which he may smelt either much or little iron. He being about to expend his skill and labour on this matter, first throws charcoal into the crucible, and sprinkles over it an iron shovelfull of crushed iron ore mixed with unslaked lime. Then he repeatedly throws on charcoal and sprinkles it with ore, and continues this until he has slowly built up a heap; it melts when the charcoal has been kindled and the fire violently stimulated by the blast of the bellows (Agricola, 1556, p. 420).*

European blast furnaces of the seventeenth century were usually square stone structures, the largest ones being more than 5 m tall and almost 2 m in diameter at the center, and until coke replaced charcoal and coal-fired steam power displaced water power, they were subject to what has been termed the dual tyranny of wood and water. Because of high preindustrial costs of bulk land transportation and the need to power furnace bellows, all larger blast furnaces had to be located not only as close to all key material inputs as practicable—the most desirable location was close both to forests (or coppiced tree plantations) and to iron ore mines—but also at sites where running water could supply sufficient kinetic energy.

And, as Biringuccio (1480–1539) reminded any would-be builder, larger furnaces should be placed into "a hillside so that the ore and charcoal can be put in easily from above on level ground, which bears the load the animals bring there, for these blast furnaces are never so small that they need less than fifty or sixty sacks of charcoal and six or eight loads of ore continually" (Biringuccio, 1540, p. 152). Molten iron accumulating at the furnace bottom was periodically discharged into sand molds forming characteristic pigs. Pig iron was converted to bar (wrought) iron in forges; these manufactories could be located closer to the market, but they were often adjacent or not too distant from blast furnaces, because they also needed a great deal of charcoal for their operation, because the refining process entailed high metal losses (up to 30%), and because the most productive forges used waterwheel-powered hammers.

Excavations of some old forge sites—be it from the Roman-era France (Domergue, 1993) or from late medieval and early modern England (Belford, 2010)—show how long-lived and how extensive some of these operations were, with what sophistication they used water power for their operations, and what a wide range of products they were manufacturing. During the fourteenth century, European cast iron found a new large market in an increasingly larger scale production of field guns (cannons) and cannonballs (displacing the traditional stone shots), but most pig iron continued to be converted laboriously into wrought iron.

CHARCOAL

Charcoal was the only source of carbon used to reduce ores in preindustrial societies (Biringuccio, 1540; Porter, 1924). Thermodynamics explains why the fuel was produced by pyrolysis (heating in the absence of oxygen) already in prehistoric times, thousands of years ago: thermochemical equilibrium calculations indicate that carbon is a preferred product of pyrolysis at moderate temperatures, with water, CO_2, CH_4, and traces of CO as byproducts (Antal & Grønli, 2003). Traditional charcoaling producing the fuel inside simple earthen kilns (wood covered by mud or turf) was very inefficient, with only 15–25% of the mass of the dry charged wood converted to charcoal; in volume terms more than 20 m^3, and only rarely less than 10 m^3 of wood were consumed to produce a tonne of charcoal (Fig. 1.1). Similarly, iron smelting during the antiquity was also extraordinarily inefficient and medieval practices brought only marginal improvements. Johannsen (1953) estimated that medieval bloomery hearths required at least 3.6 times, and as much as 8.8 times, weight of fuel than the mass of ore.

This meant that even when working with high-grade iron ores medieval bloomeries would consume commonly no less than 10 and up to 20 units of charcoal per unit of hot metal. For comparison, by 1800 a typical ratio was down to 8:1, a century later the best practices required just 1.2:1, and the late nineteenth-century Swedish furnaces needed just 0.77 kg of charcoal for every kg of hot metal (Campbell, 1907; Greenwood, 1907). Average 1:4 charcoal:wood ratio meant that (with respective energy

Figure 1.1 Charcoal making in early seventeenth-century England depicted in John Evelyn's *Sylva* (1664).

densities of 30 and 18 MJ/kg) the traditional charcoaling process lost about 60% of input energy. Charcoal's energy density of 30 MJ/kg is almost exactly twice that of air-dry wood and 20–30% higher than energy density of good bituminous coals. Free-burning charcoal could reach temperatures of 900°C, and with forced air supply (initially by using hollow wooden or bamboo tubes, later providing more powerful blast by waterwheel-powered bellows) its combustion could deliver nearly 2000°C, far higher than the melting points of common metals (lead melts at 232°C, copper at 1083°C, and iron at 1535°C).

Although charcoal can be made from reeds and plant stalks, such sources would be entirely inadequate even for small blast furnaces and their operation depended on cutting down natural forests or establishing and harvesting coppicing tree plantations. Production of charcoal (there was no difference in preparing the fuel for metallurgical and other uses) is described in many old and modern sources (Fell, 1908; Greenpeace, 2013; Uhlmann & Heinrich, 1987). English coppicing allowed 15–25 years of growth between harvests. Cut wood for charcoaling was arranged concentrically in circular piles; it was covered by coarse grass and rushes and a thin layer of fine marl or clay. Charcoaling lasted a full day, and after the pile cooled the cover was raked off and charcoal was gathered for transportation to the nearest furnaces.

Production of charcoal for small bloomeries was not a constraint on metal production in forested regions or in areas where coppicing of relatively fast growing trees provided adequate wood supply, but centuries of metal smelting in semiarid and arid regions of Europe and China were a major contributing factor to deforestation. Scale of charcoal demand had changed with the diffusion of blast furnaces. Although all medieval and early modern blast furnaces remained fairly small (maximum height of just above 5 m and square cross-sections creating internal volumes of less than 15 m^3), their campaigns (periods of constant operation) lasted for many months and, particularly in regions where they were clustered, they required unprecedented amounts of charcoal. As a result, in many regions demand for charcoal was the leading cause of deforestation, and one of the best documented examples of this distress was Sussex, now among the most affluent English counties.

By the middle of the sixteenth century, every iron mill in the Weald (area between the North and the South Downs in southeastern England) consumed every year at least 1500 loads of wood for metallurgical charcoal, and the county's inhabitants petitioned the king, afraid how many towns were like to decay if the smelting continued, and how the growing scarcity and cost of wood might affect their ability to build houses, water mills and windmills, bridges, ships, and a long list of other items (from wagon wheels to barrels and bowls, which were indispensable in what was still largely a wooden society) (Straker, 1969). In 1548, the King's commission was set up to examine the impact of the Wealden iron, but that did nothing to stop the smelting: as King (2005) showed, by 1590 the Weald's charcoal furnaces produced 3.5 times as much iron as they did in 1548, but that was their peak output followed by a steady decline; by 1660 the Weald made less iron than in 1550 as the smelting shifted to other parts of the kingdom.

During the closing decades of the seventeenth century, English ironmaking campaigns lasted for 8 months (between October and May, not only due to water flow restrictions but also due to the need for annual repairs of furnace linings), and (a conservative mean) with 8 kg of charcoal per kilogram of pig iron and with 5 kg of wood per kg of charcoal, a typical furnace producing 300 t of metal would consume about some 12,000 t of wood (Hyde, 1977). That wood was harvested mostly from coppiced hardwood trees cut every 10–20 years and producing 5 t/ha; a single furnace would have thus required about 2400 ha of coppiced hardwoods (a square measuring roughly 5 × 5 km) for its annual campaign.

Efficiency of converting cast iron to wrought (bar) iron was low, with at least a third of the metal wasted in the process, and the best available information indicates that at least 32 t of wood were needed to produce a tonne of bar iron (Hammersley, 1973). Consequently, a blast furnace and an adjacent bar forge would have consumed no less than about 10,000 t of wood a year—and in the early seventeenth century, with lower smelting and forging efficiencies, it could have been twice as much. With the wood coming from coppiced tree growth harvested in 20-year rotations and yielding 5 m^3/ha, that would imply (given wood density of 0.75 g/cm^3) about 3.75 t/ha, and hence a late seventeenth-century blast furnace and a forge would have claimed annually wood from about 2700 ha. In 1700, British output of about 12,000 t of bar iron required roughly 400,000 t of charcoaling wood, or at least 100,000 ha of coppiced trees, a square with sides of nearly 32 km (Smil, 1994).

As the numbers of European blast furnaces were increasing steadily during the sixteenth and seventeenth centuries, two improvements, taller stacks and better bellows (whose sides were made of bull hides fastened to wooden tops and bottoms, and double bellows actuated by the cams on waterwheel axles) became evident (Fell, 1908). Smelting and forging of iron became gradually less wasteful and specific charcoal consumption kept on declining (Harris, 1988), but even so the charcoal use exerted another important limitation on the output of blast furnaces: its relatively low compression strength limited their height and hence their maximum capacity; moreover, the maximum blast power was restricted by the capacity of waterwheels, as well as by their low seasonal performance. Both of these limits were overcome almost at the same time during the latter half of the eighteenth century, the first one by the deployment of more efficient steam engines, and the second one by substituting coke for charcoal, the most important innovation since the beginning of iron smelting (see the next chapter for details).

PREMODERN STEEL

As already noted, Japanese *tatara* offered a direct route to steel when operated in *kera-oshi* mode. The furnaces were charged with *masa* iron sand (with low titanium content and a low melting point) which was added first, covered with charcoal, and subjected to 3 days of smelting during which more *masa* was added to produce larger pieces of *kera*. Typical output of 2.8 t of *kera* per heat (and also 0.8 t of pig iron) from 13 t of iron *masa* and about 13 t of charcoal represents only 28% metal yield. The best chunks of the produced steel—*tama-hagane* or jewel steel, adding up to usually less than 1 t per heat—were used to make the famous *nihonto*, Japanese swords. Limited output of the directly produced steel and very expensive production of steel objects had largely restricted the use of that high-quality alloy in premodern Japan to a few categories of weapons.

Although simple unalloyed iron was usually the only product of European bloomery smelting, on some occasions (with higher temperatures and longer smelting periods) the process produced small unagglomerated fragments (gromps) of high-carbon steel, and gromps were sometimes present even as components of agglomerated blooms. This minuscule steel production was simply an accidental coincidence of localized high temperature and right reducing conditions, not a desired outcome of bloomer smelting. By the beginning of the seventeenth century, the most common route to high-quality steel in preindustrial Europe was the conversion of wrought iron through carburization (cementation) that boosted its very low carbon content.

European carburization involved prolonged heating of the metal (Swedish low-phosphorus iron was a preferred choice in England) with charcoal in stone chests. Without any subsequent forging, the gradual inward diffusion of carbon into wrought iron produced a thin steel layer enveloping the core of softer iron. Producing steel by carburization of iron can impart as much as 2% of carbon by solid-state diffusion, but it is necessarily a very laborious process: it might take about 50 h to enrich the metal to the depth of 4 mm at 925°C (Godfrey & van Nie, 2004). This technique produced material perfectly suited for plowshares or for body armor. Repeated forging of carburized metal resulted in a fairly even distribution of the absorbed carbon and produced steel suited for swords.

Carburization in crucibles was also the only way to make high-quality Indian steel, but its high cost limited the metal's use to expensive, high-prestige objects, above all to swords. The best ones of these became known as Damascus steel blades, distinguished by their beautiful surface pattern of light-etched swirls on a nearly black background. Manufacturing of these blades reached the highest quality between the sixteenth and eighteenth centuries, and the technique then became virtually extinct during the nineteenth century. We now know that the metal was never produced in Syria but that it was imported from India in the form known in the West as wootz steel, a garbled transcription of *urukku* or *urukke*, the words for steel in several South Indian languages (Biswas, 1994). The quality of this metal has been admired for 2 millennia and its production and properties have fascinated metallurgists and historians for more than 2 centuries (Egerton, 1896; Feuerbach, 2006; Figiel, 1991; Mushet, 1804; Verhoeven, 1987, 2001; Voysey, 1832).

This ancient Indian method of steelmaking, with the most likely origin in Tamil Nadu, was described in great detail by Richard Burton in *The Book of the Sword*:

> *About a pound weight of malleable iron, made from magnetic ore, is placed, minutely broken and moistened, in a crucible of refractory clay, together with finely chopped pieces of wood* Cassia auriculata. *It is packed without flux. The open pots are then covered with the green leaves of the* Asclepias gigantea *or the* Convolvulus lanifolius, *and the tops are coated over with wet clay, which is sun-dried to hardness. Charcoal will not do as a substitute for the green twigs. Some two dozen of these cupels or crucibles are disposed archways at the bottom of a furnace, whose blast is managed with bellows of bullock's hide. The fuel is composed mostly of charcoal and of sun-dried brattis or cow-chips. After two or three hours' smelting the cooled crucibles are broken up, when the regulus appears in the shape and size of half an egg (Burton, 1884, p. 111).*

After removal from crucibles small ingots were shaped into bars for trading, and although a single heat yielded just enough steel for two sword blades, wootz production in some Indian regions (Lahore, Amritsar, Agra, Jaipur, Mysore, Malabar, Golconda) was carried out on an almost industrial scale so that it could supply a steady stream of metal for exports to Persia and the Turkish Empire. Correct metallurgical classification of wootz steel is a hypereutectoid ferrocarbon alloy with spheriodized carbides and carbon content of 1.2–1.8%. Relatively high carbon and phosphorus content make wootz steel brittle in soft condition but well-made swords are ductile and not prone to breaking in a battle although they may bend on impact (Perttula, 2004). Studies of actual sword blades confirmed the material's superplasticity and high impact hardness, although the latter quality was inferior compared to modern steel: the most often cited examination of Damascene swords showed Brinell hardness of 171–264 compared to 313–473 for Solingen steel (Zschokke, 1924).

In order to get its fine grain and plasticity, wootz steel must be forged in a narrow range of 850–950°C, well below the white heat of 1200°C that would make the metal brittle (Biswas, 1994). About 50 cycles of forging may be needed to form the final blade shape from a wootz ingot.

Modern metallurgical analyses and reconstructed wootz steel production demonstrated that the metal's characteristic band formation resulted from microsegregation of small amounts of carbide-forming trace elements, above all vanadium, molybdenum, and manganese (Verhoeven, Pendray, & Dauksch, 1998; Verhoeven, 2001). Forging results in tiny (6–9 μm in diameter) articles of cementite (iron carbide, Fe_3C) concentrated in clustered bands between 30 and 70 μm apart and after etching the surface with acid they show as delicate, irregularly wavy white lines within dark steel matrix. Ladder-like undulations and rose patterns displayed in some swords are created by cutting grooves and drilling holes into the surface of a blade and then forging it into its final form.

Grazzi et al. (2011) used a noninvasive analysis (thermal neutron diffraction) to compare ancient and historic steels from Japan, India, and Europe and their findings confirmed the known compositional differences caused by specific production processes. European metal (from crossbow, arrow, pole arm, and gun) required the least amount of fuel to smelt and it is dominated by ferrite (92.8–97.8%) with no observed cementite and martensite; very expensive, high-carbon (>1% C) Indian objects (sword, shield, and knives) had between 70–85% of ferrite and 9–27% of cementite (but no martensite); and the composition of Japanese swords gave intermediate values (and it had a similar position in terms of fuel demand): much like the European metal objects it is dominated by ferrite (roughly 91–96%) but with significant presence of cementite (about 1.5–4.5%) and with traces of martensite (0.3–0.9%).

In contrast to European and Indian practices, traditional Chinese metallurgists, able to produce carbon-rich liquid iron, faced the opposite challenge in steel production as they lowered high-carbon content through decarburization, and ever since the eventual worldwide adoption of blast furnaces producing liquid iron all modern steelmaking processes (first Bessemer converters, then open-hearth furnaces, now basic oxygen furnaces) are relatively rapid and highly effective forms of the process. The two methods of Chinese decarburization were by blowing oxidizing blasts of cold air over cast iron (known as the hundred refinings), and what Needham (1964) called co-fusion, immersion of wrought iron in molten cast iron practiced at least since the middle of the first millennium of the common era.

We do not know how this method was eventually mastered in Europe (by transfer from China or by independent discovery) but Biringuccio's

(1540, p. 69) description of the immersion and subsequent cooling makes it clear that this was one of the most admirable feats of traditional ferrous metallurgy:

> *This bath is called "the art of iron" by the masters of this art ... they keep in it this melted material with a hot fire for four or six hours, often stirring it up ... When they find that it has arrived at the desired point of perfection they take out the lumps ... cut each one in six or eight small pieces. Then they return them to the same bath to heat again ... at last, when these pieces are very hot, they are taken out and made into bars ... after this, while they are still hot ... they are suddenly thrown into a current of water as cold as possible... In this way the steel takes on that hardness which is commonly called temper; and thus it is transformed into a material that scarcely resembles what it was before.*

Medium- or high-carbon steel was also produced directly in small quantities in parts of sub-Saharan Africa and in Sri Lanka. The simplest ancient technique that could directly produce good-quality medium-carbon steel was practiced in East Africa since the first centuries of the common era. Skilled steelmakers built circular, cone-shaped mud furnaces up to 2m tall and fueled them with charcoal laid over a pit of charred grass. Schmidt and Avery (1978) described a re-enactment of this traditional procedure during which eight men operating goatskin bellows connected to ceramic tuyères were able to raise the temperature above the melting point of iron.

Africa's carbon-containing blooms were made possible either by deep insertion of long tuyères or by building tall furnaces whose height maximized naturally induced draft (Avery & Schmidt, 1979; Schmidt & Childs, 1995). Long tuyères extending deep into the bowl resulted in preheating of blast air and hence in higher temperatures (1300–1500°C) than could be achieved in European blast furnaces relying on cold blast. In Sri Lanka high-carbon steel was produced by building the furnaces on the western sides of hills and ridges exposed to strong seasonal monsoonal winds. Experiments demonstrated that strong natural draft created by these flows helped to produce heterogeneous but sufficiently extensive carburization to indicate that the desired product of this high-wind smelting was a metal that can be classed as high-carbon steel (Juleff, 1996; 2009).

But throughout sub-Saharan Africa, the traditional artisanal production remained small scale and localized, and iron artifacts remained relatively rare before the European imports became available. Some Sri Lankan metal was exported during the early medieval era to make Middle Eastern swords, but in that society, too, daily artifacts continued to be made of wood and clay and iron objects did not become common until the nineteenth century. Iron use was much more common in parts of ancient Europe, but finished objects made of high-carbon steel (1.5–2.1% C) are rare before the seventh century CE. For example, Godfrey and van Nie (2004) analyzed a Germanic ultrahigh carbon (about 2%) steel punch of the Late Roman Iron Age in the Eastern Netherlands that was most likely produced in solid state from thin strips of metal that were carburized separately and then welded.

Iron objects in Western Europe did not become more common until the eleventh century, when many plows got iron shares (or at least iron colters), and by the thirteenth century metal tools and parts for construction and shipbuilding were used more frequently (Crossley, 1981). But small iron manufactures, so ubiquitous in later eras and so unremarkable today, were still in short supply: individual metal cutlery was absent as "people served themselves from the common plate ... everyone drank from a single cup ... knives and spoons were passed from person to person" (Flandrin, 1989, pp. 265–266)—and there were no small table forks.

This began to change during the seventeenth century when iron objects and components also became a more prominent part of European proto-industrialization. Biringuccio's book, published at the beginning of the early modern era, contains a paragraph-long list of iron products that required specific smithing expertise, ranging from such massive items as anchors, anvils, and guns and such agricultural implements as plowshares, spades, and hoes to "more genteel irons such as knives, daggers, swords" as well as personal armor, gouges, drills, locks, and keys and "many more ... things that are made or can be made of iron" (Biringuccio, 1540, p. 370). Their number was to grow rapidly during the eighteenth and the early nineteenth century as it became easier to produce iron in larger quantities and with lower costs: by 1850, although steel was still expensive and relatively rare, iron was ascendant.

CHAPTER 2

Rise of Modern Ferrous Metallurgy, 1700–1850

Coke, Blast Furnaces, and Expensive Steel

During the first decades of the eighteenth century standard metallurgical setups (blast furnaces located near iron ore deposits and close to wooded areas with streams), procedures (charcoal-fueled smelting producing pig iron, most of which got converted to wrought iron in forges), and products (cast iron, wrought iron, small shares of steel) differed little from practices that prevailed for most of the seventeenth century (Fig. 2.1). Specifically, there was nothing unique about the practices of the British iron industry except for the fact that the country had to rely more on the imports of the Swedish iron. In many countries this pattern of pig iron production persevered for another century—but by 1800 the British iron industry was the world's undisputed leader thanks to a combination of technical advances that made it more productive and more efficient and brought unprecedented expansion of aggregate output. The first half of the nineteenth century brought fewer major innovations, but the key technical advance of the period, the introduction and widespread adoption of hot blast, had tremendous impacts on both the efficiency and productivity of blast furnaces.

Evans, Jackson, and Rydén (2002) correctly note that the term *iron industry* is anachronistic when describing the activities of the eighteenth century: the proper English term was *iron trade*, a chain of activities stretching from smelting to manufacturing of iron items and involving labor ranging from forge masters and slitting mill proprietors to international merchants and artisanal workers who turned the metal into an increasing variety of products. Only a small share of pig iron was used for direct casting to produce cannons, cannonballs, and shot as well as cooking pots; most of the pig iron was converted by forging into bar iron that was either turned into nails and spikes or that was fashioned by smiths into a wide range of tools and implements including such ubiquitous objects as hinges, locks, springs, knives, sickles, and scythes.

Still the Iron Age.
DOI: http://dx.doi.org/10.1016/B978-0-12-804233-5.00002-6

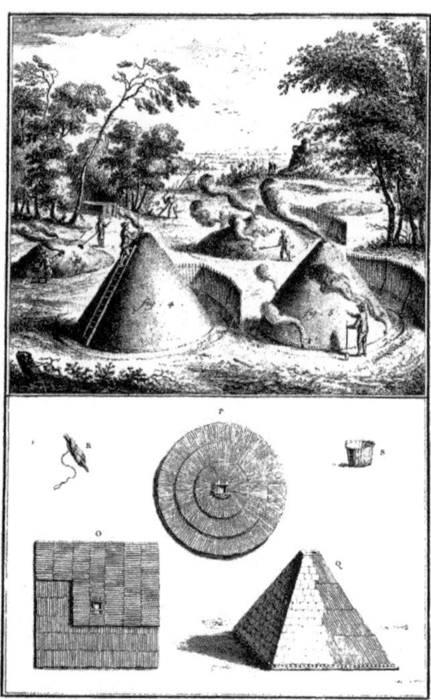

Figure 2.1 Steps in charcoalmaking illustrated in *L'Encyclopédie ou Dictionnaire Raisonné des Sciences des Artes and des Métiers* (1751–1780).

EUROPEAN AND BRITISH IRONMAKING BEFORE 1750

We have no reliable Chinese or Indian production data for the eighteenth century, but European aggregates are good enough to state with certainty that Sweden and Russia were the leading producers of pig iron, and that those two nations were also dominant exporters of bar iron, and that England and Wales were its leading importers. Competitively priced Swedish exports went initially through Danzig, after 1620 they were destined mostly for Dutch ports, and since the middle of the seventeenth century England became their leading buyer (Evans, Jackson, & Rydén, 2002). English imports reached nearly 14,000 t in 1700 when they amounted to about 80% of all bar iron brought to England, rose to just over 20,000 t by 1735, and then until the 1750s were close to the total domestic output of roughly 20,000 t of iron a year (Harris, 1988). Total Swedish exports averaged 42,500 t/year during the 1740s and they rose to 48,000 during the 1790s, still larger than the iron output in England and Wales.

But by that time Russia replaced Sweden as the principal exporter, supplying nearly two-thirds of all imported bar iron during the 1780s, with Swedish shipments retaining high-quality market in Sheffield (Rydén & Ågren, 1993). During the second half of the seventeenth century Russia was an importer of Swedish iron, but this dependence was rapidly eliminated by the construction of large ironworks in the middle Urals region (with Nizhny Tagil as the center) during the reign of Peter I. Small-scale iron production by peasants was present in the region for decades, but between 1701 and 1730 33 large ironworks (13 of them state-owned) were completed, setting the foundations for Russia's emergence as the world's largest producer of pig iron later in the century (Minenko et al., 1993).

The best iron came from the works owned by the Demidov family. Their ironmaking began with Nikita Demidov (1656–1725) in Tula. It then moved to the Urals where it was greatly expanded by his son Akinfiy Nikitich Demidov (1678–1745). The works were then inherited by his grandson Nikita Akinfyevich Demidov (1724–1789), and their brand of cast iron (*Staryi sobol*, Old Sable) was preferred by the English importers (Hudson, 1986). Cast iron output of the Urals region rose from less than 10,000t in 1725 to nearly 123,000t in 1800, and Russia's total pig iron output during the last decade of the eighteenth century was just over 200,000t a year, at least 20% ahead of the UK production (King, 2005). Aggregate production data for non-European countries are just estimates, but it appears that in 1750 both the Chinese and the Indian cast iron productions were similar to the Russian smelting, and that the worldwide production of liquid iron was on the order of 800,000t (Pacey, 1992).

Substantial exports of Russian bar iron to the United Kingdom started before 1720 and peaked before the century's end: in 1794 Russia sold 63,600t of iron bar abroad (Minenko et al., 1993). The best reconstruction of iron trade in early modern England and Wales shows three great reversals (King, 2005). In 1500 domestic output was only about 20% of total consumption; it surpassed imports of bar iron, mainly from Spain, by 1550, and a century later it accounted for more than 80% of the total use. But by 1690 exports, overwhelmingly from Sweden, surpassed the domestic output and until the time they peaked (at nearly 46,000t in 1770) they supplied between half and two-thirds of total consumption. Imports from Sweden dominated until 1760, and the imports of Russian iron from the Urals (first moved by boats on rivers and canals to Sankt Peterburg) peaked during the 1770s when England covered two-thirds of its iron demand by foreign metal (King, 2005). The fame of Sheffield steel products rested to a large extent on pig iron from the Urals.

Reconstruction of English pig iron production shows nearly two centuries of stagnating or only very slowly rising output: annual rates approached 10,000 t before the end of the sixteenth century, were only marginally higher a century later, and were no higher than 25,000 t by 1750. Only the elimination of the charcoal ceiling on English pig iron production brought a rapid elimination of imports (falling from nearly 50% in 1795 to less than 10% by 1810) and massive expansion of iron output (at 2.5 Mt in 1850 it was 100 times higher than in 1750). And there were two other factors besides switching from charcoal to coke (first used in 1709, but extensive conversion only after 1750): switching from water to steam power, beginning in 1776 when the first steam engines patented by James Watt (1736–1819) began commercial operation (Fig. 2.2), and adoption of coal-fired refining of the metal, beginning in 1784 when Henry Cort (1741–1800) got his patent for the puddling and rolling process.

Reliance on two renewable flows of energy—on wood-based charcoal and stream flows—imposed fundamental restrictions on capacities and productivities of early eighteenth-century blast furnace operations (Fig. 2.3). Although the efficiency of iron smelting kept on improving, overall capacities of charcoal-based metallurgy remained inherently limited by the access to wood, and capacities of individual furnaces were limited by charcoal's structural properties. Water flows were the other key restrictions: annual smelting campaigns in Britain were limited to about 30 weeks between October and May because summer water flows were commonly inadequate to generate power needed for the bellows. Moreover, even with full stream flows the maximum power of commonly deployed early eighteenth-century wheels (with diameters up to 7 m in 1700 and 12 m by 1750, and with power ratings of up to 7 kW, an equivalent of more than nine horses, by the mid-eighteenth century) could not provide optimum blast for large charcoal-fueled furnaces, and the use of coke needed even stronger blast.

During the eighteenth century ore smelting faced no wood limits in richly forested Sweden or Russia and hence their charcoal-fueled iron output kept on rising, but availability of suitable wood was a limiting factor in further expansion of English iron industry. Hammersley (1973) estimated that the maximum countrywide harvest of wood for charcoal would have been on the order of 1 Mt/year but the actual demand never

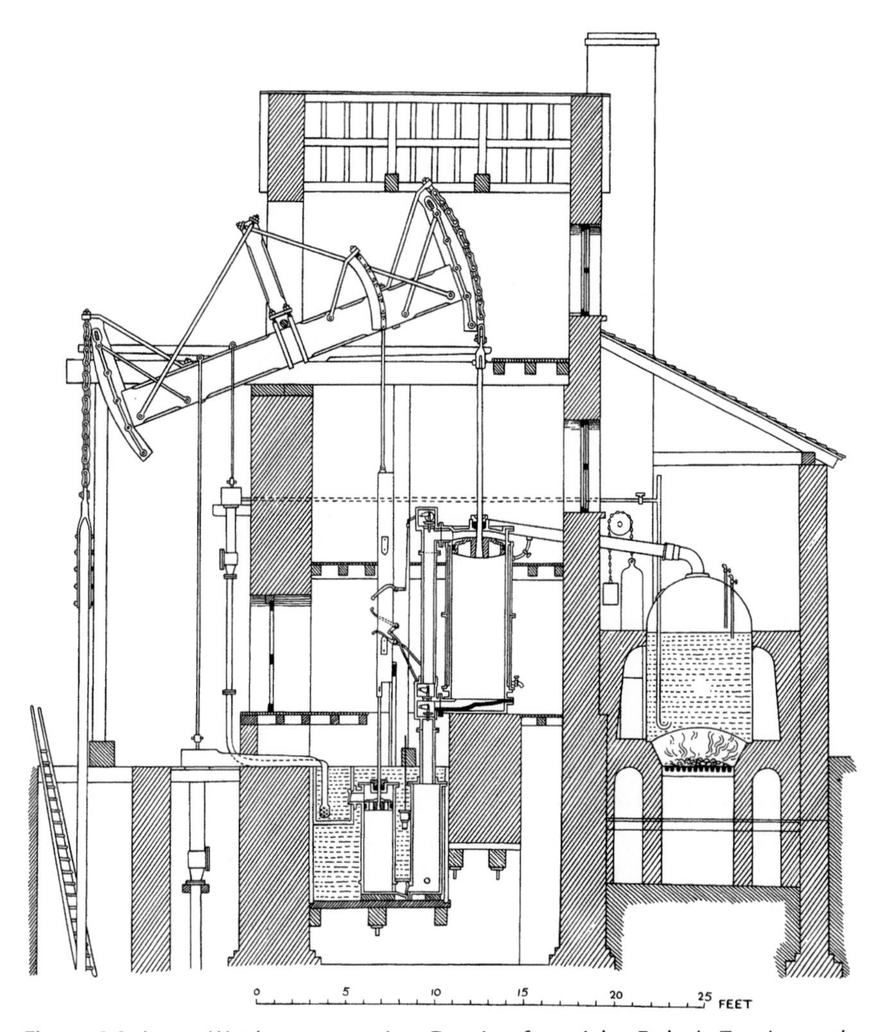

Figure 2.2 James Watt's steam engine. Drawing from John Farley's *Treatise on the Steam Engine* (1827).

surpassed that rate because English and Welsh ironmasters were producing annually less metal during the first four decades of the eighteenth century than they did between 1600 and 1640 (King, 2005).

By 1720 the annual output of 60 British furnaces reached 17,000 tonnes of pig iron and (with about 40 kg of wood per kg of metal) it required about 680,000 tonnes of charcoaling wood. An additional 150,000 t of wood were needed to forge iron bars, and 830,000 t of wood

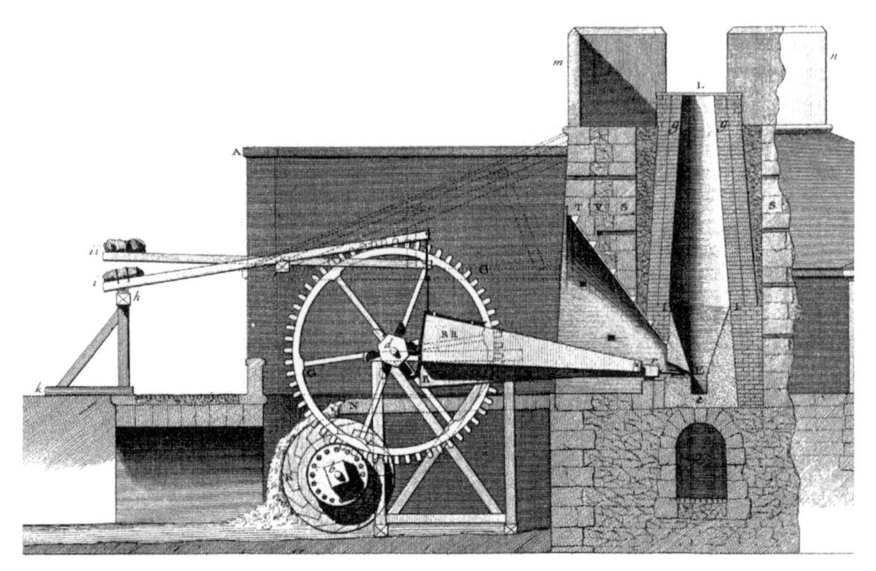

Figure 2.3 Cross-section of an eighteenth-century blast furnace. Engraving from *L'Encyclopédie ou Dictionnaire Raisonné des Sciences des Artes and des Métiers* (1751–1780).

coming from coppicing would have required nearly 1,700 km^2 of trees, an equivalent of a square with sides of almost 41 km. Such a demand could have been supported by properly managed plantings in perpetuity and, indeed, English charcoal prices remained steady during the first half of the eighteenth century. But, as I have just shown, that was possible only because the domestic charcoal was fueling less than half of total metal consumption as imports from Sweden and Russia became more prominent.

Charcoal is an excellent fuel: its energy content is almost identical to that of coke and hence it generates heat required for blast furnace reactions; its combustion releases CO, the gas that reduces iron oxides to hot metal. But besides generating heat and reducing gas, charcoal and coke also have key physical functions in blast furnace operation, to support the burden column and to create permeability that allows the ascent of heat and reducing gases and downward flow of slag and metal; fundamentally, a blast furnace is a counter-current reactor. But commonly used metallurgical charcoal cannot support heavy burdens because it is a rather friable material that could not maintain open spaces under heavier loads of ore and limestone and would get eventually crushed to dust, making iron smelting impossible.

There are many modern methods used to measure the strength of coke at different temperatures (Yamazaki, 2012) but for the most relevant comparison with charcoal we have to look at the compressive strength of the two fuels: good charcoal made from solid wood has compressive strength of about 4 MPa compared to 15 MPa for typical metallurgical coke at 1000°C, with the value declining to about 12 MPa at 1600°C (Emmerich & Luengo, 1996; Haapakangas et al., 2011). This difference explains why the height of charcoal-fueled blast furnaces was not more than about 8 m, with annual metal output averaging only about 300 t (and exceptionally about 700 t) per furnace (Sexton, 1897). Consequently, the only way to increase aggregate output was to build more furnaces.

BRITISH TRANSITION TO COKE

Adoption of metallurgical coke for iron smelting was definitely one of the greatest technical innovations of the modern era as it severed the dependence on wood, opened the way toward a huge growth of furnace capacities and to multiplication of annual outputs, and freed smelting locations from the proximity to streams able to power furnace bellows. There were several reasons why the replacement of charcoal by coke in English and Welsh furnaces was a rather protracted affair. Initially it was the easy British access to affordable imports of the Baltic iron (although the Russian smelting was done largely in the Urals, the shipments came via Sankt Peterburg), and the first commercial uses of the fuel indicated that coke-based smelting was not financially attractive (Harris, 1988; Hyde, 1977).

Coke was first used in England during the early 1640s for drying malt (a task that could not be done with coal as its combustion produced copious particulate and sulfur emissions), and unsuccessful attempts at its use (and also of coal and peat) in metal smelting took place during the latter half of the seventeenth century, but it was only in 1709 when Abraham Darby (1678–1717) became the lone pioneer of iron ore smelting with coke. Hyde (1977) offered a convincing explanation why English ironmasters of the first half of the eighteenth century did not follow Darby's example (his two furnaces in Coalbrookdale used coke exclusively after 1720, and one in Wiley used only coke since 1733) before the early 1750s.

Although some 25 charcoal-fueled furnaces were closed between 1720 and 1755, the aggregate output of charcoal-smelted iron rose from nearly 19,000 to almost 25,000 t during the intervening 35 years. The reason was neither any secrecy surrounding Darby's innovation nor an inferior quality of coke-smelted iron, but significantly higher operating costs of coke-fueled furnaces and no major cost difference in capital cost of new furnaces. Hyde (1977) calculated that operating costs of the two processes may have become equal by the late 1730s, but because of the large amount of coke consumed the overall costs were in favor of charcoal furnaces until the early 1750s.

Darby and his successors were able to make the coke-based smelting profitable "*in spite of higher costs of the new process* because they received higher than average revenues from a new by-product of coke pig iron—thin-walled castings" (Hyde, 1977, 40). This technique, patented in 1707 before Darby began his coke smelting, benefited from higher fluidity of Si-rich coke-smelted iron that could be used to produce much thinner pots (with half as much mass as those made of the charcoal-smelted iron) with fewer defects. Moreover, Hyde (1977) also concluded that making bar iron from coke pig iron was more expensive than making it from charcoal cast iron because the former liquid metal contained more silicon.

King (2011) revisited Hyde's (1977) explanations and his detailed examination of Coalbrookdale business records (extant in four account books) and confirmed the conclusion regarding the costs of pig iron smelted with coke, but found that the same argument did not apply to the production of bar iron. Account books show enormous coke consumption in Coalbrookdale furnaces during the 1720s and its gradual decline during the 1730s. But the accounts, and comparisons with other forges, show that the poor performance of Coalbrookdale enterprise was not due to inherent problems with coke-smelted pig iron but rather to a demonstrable fact that it was a small and inefficiently run enterprise.

Delay in widespread coke adoption was thus largely a matter of bar iron price: "Whatever technical difficulties existed in the use of coke pig iron in forges in the early 1720s, these were evidently overcome by the end of that decade, but the depressed state of the iron trade discouraged the introduction to the market of coke-smelted forge pig iron, until the industry benefited from an economic upturn in the 1750s. That upturn can in part be attributed to the Swedish limitation on their iron production, which began a few years earlier" (King, 2011, 154). English producers responded almost immediately by building new coke-fueled furnaces after the mid-1750s. Nearly 30 coke-based furnaces were built between 1750 and 1770, and their share of pig iron output rose from just 10% to 46% (King, 2005).

This was an epochal change, from the dependence on a resource that was renewable but already in short supply in many regions and whose maximum realistic exploitation could not support the future expansion of iron production to the dependence on a nonrenewable fuel that could be produced inexpensively from abundant coal deposits and whose output could be scaled up to meet any foreseeable expansion of iron industry. And the substitution removed the pressure on continental forests: Madureira (2012) calculated that in 1820 52% of Belgium's forested area was used to produce metallurgical charcoal, and that even in much larger and much more forested France and Sweden the shares were about 15% by 1840.

Impossibility of long-term reliance on charcoal is easily illustrated with relevant calculations for the exceptionally wood-rich United States. US nationwide iron output statistics began in 1810 when the smelting of 49,000 t of pig iron consumed (assuming an average rate of 5 kg charcoal or least 20 kg of wood per kg of hot metal) about 1 Mt of wood. Even if all that wood would have come from natural old-growth hardwood forests storing around 250 t/ha (Brown, Schroder, & Birdsey, 1997), and even if all above-ground phytomass were used in charcoaling, an area of nearly 4000 km^2 (a square with a side of almost 63 km) would have to be cleared every year to sustain that level of production. Rich US forests could support an even higher rate and by 1840 all US iron was still smelted with charcoal, but after a subsequent rapid switch to coke energized nearly 90% of iron production by 1880 and future increases in iron production could not be based on charcoal, in 1910—with iron output at 25 Mt and even with much reduced charges of 1.2 kg of charcoal and 5 kg of wood per kg of hot metal—the country would have required 125 Mt of wood a year.

That requirement alone (leaving aside all charcoal need for further metal processing) would have necessitated (even with high average increment of 7 t/ha in natural forests) annual wood harvest from nearly 180,000 km^2 of forest (Smil, 1994). That area would be equal to Missouri or Oklahoma (or a third of France), and if it were a square its side would go from Philadelphia to Boston, or from Paris to Frankfurt. Obviously, even forest-rich America could not do this. Moreover, there could be no doubt about its superiority as a metallurgical fuel. Coke is produced by heating suitable kinds of bituminous coals (they must have low ash and low sulfur content) in the absence of oxygen: this pyrolysis (destructive distillation) drives off virtually all volatile matter and leaves behind nearly pure carbon with low apparent density of just 0.8–1 g/cm^3 but with higher heating value at 31–32 MJ/kg, roughly twice as energy dense as air-dried wood but only slightly more energy dense than the best charcoal.

As with most technical advances, the efficiency of early coke-producing methods was very poor. For more than a century the standard way to make coke was in enclosed beehive ovens (Sexton, 1897; Washlaski, 2008). These hemispherical structures (American beehives had diameter of about 3.8 m) were usually built in banks (called batteries, with some American batteries eventually having 200 to 300 beehives), often into a hillside (making it easier to cover them with earth), and always with strong frontal retaining walls. After 4 or 5 days of preheating (first with wood, then with coal) the starter fuel was removed, front doors were bricked up to two-thirds of their height, and ovens were charged with coal. Average charge for a standard oven was 5–5.5 t, loaded coal was leveled with an iron bar, doors were bricked up and sealed with clay, and for the next 2 to 3 days (burning periods varied between 40 and 75 h) slow-burning beehive ovens lit the night skies with an orange-reddish glow and emitted hot gases through their open tops (trunnel heads).

Once the controlled burning ended coke was finished by quenching with water, doors were broken up, and the fuel was removed from beehives to be transported to blast furnaces. Early beehive coking consumed up to 2 t of coal per tonne of coke; later the yield increased to 60% and eventually to about 70%. Coke's ability to support heavier charges of ore and limestone made it possible to build taller blast furnaces with larger capacities and higher outputs, and this, in turn, increased demand for coke. Some of these early coke furnaces used Newcomen's inefficient steam engines, and after 1776 the diffusion of coke-based smelting was greatly aided by the adoption of Watt's steam engines as the drivers of more powerful bellows: they were used in that way in 1782 and by 1790 England and Wales had 83 coke-fueled furnaces in operation and 71 had steam-powered bellows (Hyde, 1977). But charcoal-fueled furnaces did not disappear: in 1810 they still smelted one-third of English and Welsh iron.

LARGER FURNACES AND HOT BLAST

Coke-based smelting using steam-driven blast severed the traditional location links to forested regions with fast-flowing streams and led to rapid increases of individual furnace volumes and hence to an unprecedented growth in pig iron production. In order to follow the subsequent design changes in their height, volume, and output capacity it is first necessary to understand the basic structural components of these increasingly massive structures. Their modern designs are always built on substantial foundations, and at their bottom is a circular hearth whose top perimeter is pierced by water-cooled tuyères blowing in pressurized and heated air and where the liquid metal and slag are collected between tappings.

Then comes the bosh, a short, truncated, and slightly outward-sloping cone whose interior contains the highest smelting temperatures. The belly, the furnace's widest section, is between the bosh and the stack (shaft), the longest, and only slightly narrowing, section where downward movement of ore, coke, and flux meets the upward movement of hot, CO-rich gases that reduce iron oxides. The furnace throat is surmounted by a top cone that contains arrangements for furnace charging as well as for the collection of hot furnace gases. Molten metal is periodically released through tapholes whose clay plugs are opened up by a special drill, and the slag (now commonly used in construction or as a fertilizer) is taken out through cinder notches. Furnace campaigns (uninterrupted operations) last many years before the smelting stops and the furnace is let to cool for relining of its walls with refractory materials and of its hearth with carbon. Furnaces' growth in height and capacity changed every one of their basic structural components.

The first coke-fueled blast furnaces were not taller (about 8 m) and were no more voluminous (with inside capacities of less than 17 m^3) than their large contemporary charcoal-fired structures, but in 1793 Neath Abbey was the first furnace to reach the exceptional height of 18 m. By 1810 coke-fueled furnaces were around 14 m tall with volumes of more than 70 m^3, and their typical annual outputs increased from about 1000 t/year during the late 1780s to 1500 t/day by 1810. The next two decades saw a few fundamental changes: by the late 1830s large British furnaces were less than 15 m tall, with stacks of less than 9 m and square hearths with sides of less than 2 m: vertical cross-section shows a narrow hearth, a relatively tall and sloping bosh, and a fairly short narrowing stack.

Lowthian Bell (1816–1904), one of the leading metallurgists and industrialists of the nineteenth century, concluded that such furnaces were too short and too narrow for efficient iron ore reduction, and in 1840 his redesign introduced a furnace that has served as a prototype of all subsequent modern designs (Bell, 1884). Bell increased the overall height by two-thirds, to 24 m, with a stack of nearly 17 m (stack:total height ratio of 0.7 compared to previous 0.6); his bosh was relatively shorter (bosh:total height ratio of 0.22 compared to previous 0.44) and less sloping, and he also enlarged the top opening by more than 80% and the hearth diameter by a third (Bell, 1884).

These larger coke-fueled furnaces opened the way to unprecedented increases of iron output and made the United Kingdom Europe's largest pig iron producer. Production in England and Wales grew 2.33 times during the last decade of the eighteenth century to 282,000 t. That was nearly four times as much as the stagnating Swedish production (King, 2005; Olsson, 2007). A decade later the British output was in excess of 400,000 t/year while the US shipments in 1810 (the first year for which we have an output estimate) reached just 49,000 t (USBC, 1975). By far the most important smelting innovation of the first half of the nineteenth century was the introduction of hot blast during the late 1820s and its nearly universal use just 15 years later.

James Beaumont Neilson (1792–1865) was not an experienced ironmaster but an outsider who was able to apply a general observation (he noticed how flame luminosity can be enhanced by supplying it with preheated air) to a specific engineering problem. Neilson patented his hot blast technique in 1828 and it was first used in the Clyde Ironworks in Scotland, and ironworks in Scotland continued to adopt hot blast much faster than their English counterparts and soon came to claim the largest share of newly installed blast furnace capacities (Birch, 1967; Harris, 1988). In retrospect this may be an eminently logical step but it was resisted by conservative ironmasters who believed in the efficacy of cold blast and maintained, erroneously, that hot blast would lower the quality of metal.

Initially, all preheating was done by burning a relatively small amount of coal outside of a furnace to heat the blast air, but this additional outside combustion resulted in impressive fuel savings inside the furnace. In 1829 the first preheating, to just 150°C, saved more than a third of fuel compared to cold blast, and in 1833 preheating to 325°C saved an additional 45% (Bone, 1928). Part of these savings was due to less wasteful coking methods and better boiler efficiencies, but Bell (1884) concluded that the savings attributable to hot blast, equivalent to at least 1.75 t of coal, were achieved by additional external combustion of 100–150 kg of coal, a return that justified the payoff as sufficiently astounding.

Besides the universal benefits of hot blast—reduced consumption of coke and limestone, increased furnace productivity (by about half as furnaces could be tapped more often), and better quality of cast iron—there were three important specific gains. Hot blast opened the way to large-scale ironmaking in Scotland by making it possible to use local low-quality iron ore (blackband ironstone discovered at the beginning of the nineteenth century) and to substitute local raw coal for more expensive coke. As a result, Scottish production of cast iron had more than quintupled during the 1830s to nearly 200,000 t/year. Similarly, hot blast in South Wales allowed the use of local anthracite, and the use of high-quality Pennsylvania anthracite (for the first time by Frederick W. Gessenhainer in 1836), became the foundation of the state's large iron industry (Bone, 1928). Hot blast had also changed the access to the hearth: the furnace's outer casing was supported by cast iron pillars that allowed access from every direction and made it possible to deploy a larger number of better (water-cooled) tuyères that could withstand the constant heat of more than 300°C.

The combination of hot blast and larger furnaces led to another round of productivity increases: 15-m tall cold-blast furnaces of the late 1820s produced less than 80 t of metal per week; two decades later 18-m tall hot-blast furnaces produced up to 200 t/week (Birch, 1967). The next logical step took another two decades to accomplish: why to heat gases in blast stoves when large volumes of hot gases were constantly escaping from the open tops of blast furnaces? This large loss of heat began to be tackled during the late 1840s (James P. Budd filed his heat recovery patent in 1845), and soon afterwards furnaces became closed, usually by a massive cup (fixed) and cone (movable) apparatus that was introduced by George Parry (1813–1873) in 1850. The final step in eliminating furnace waste heat came with the adoption of regenerative hot-blast stoves during the 1860s (see the next chapter).

As productivities kept on increasing the output per furnace rose up to 2500 t/year during the 1820s and the maxima surpassed 10,000 t/year during the 1830s. Aggregate British output of pig iron rose by about 70% during the 1820s (to 677,000 t); it then doubled during the next decade (to 1.396 Mt), and by 1850 it was up by almost 80% to 2.5 Mt produced by 655 blast furnaces compared to just over 300 furnaces in 1825 when the production was less than a quarter of the 1850 level (Birch, 1967). British lead became enormous when compared to any other European country as well as to the United States: by 1850 Swedish output of pig iron was about 125,000 t, and although American pig iron shipments rose from about 49,000 t in 1810 to 150,000 t in 1830 and 512,000 t in 1850, the latter total was less than a quarter of the British output (USBC, 1975), and it took the United States almost another four decades to become the world's largest producer of iron.

WROUGHT IRON

A higher share of the metal produced in coke-fueled furnaces was cast to produce parts for machinery and construction, and higher liquidity of that metal made it possible to produce more delicate castings. At the same time, converting pig iron into wrought iron, previously done in forges associated with furnaces, became easier with the introduction of Henry Cort's puddling process patented in 1784. This decarburization of iron was achieved by continuous stirring (or more correctly turning and pushing) of molten pig iron with long rods (rakes) through the doors of reverberatory furnaces (fired with coke) in order to decarburize it (by exposing it to oxygen) and producing a nearly pure metal containing less than 0.1% of carbon.

The method was widely adopted in the United Kingdom after the mid-1790s, and it became the dominant means of decarburization during the first half of the nineteenth century: British wrought iron production was surpassed by steel production only in 1855. Puddling's yield was greatly improved by Joseph Hall (1789–1862), who substituted sand at the furnace bottom (silica furnace, commonly known as Cort's dry puddling) by iron oxide, a change that produced partially liquid iron (hence "wet" puddling), yielded much less slag, and allowed a more complete conversion of pig to malleable iron: the losses were less than 15% and no more than 25% of the charged metal compared to at least 30% of pig iron that was incorporated in sandy slag. This has been often portrayed as an improvement of Cort's process, but, as Flemings and Ragone (2009) point out, Hall's innovation was entirely different from a metallurgical point of view because it involved oxidation of carbon by iron oxide rather than by oxygen present in the furnace atmosphere. Hall's experiments began in 1811 and his process was commercialized during the 1830s.

The sequence began with the firing of a reverberatory furnace which was lined with mostly Fe_3O_4; between 250–175 kg of cast iron, and some iron oxide, were shovelled in, the openings were closed, and the furnace was heated for half an hour to melt the metal. Then the puddler began to stir, exposing the metal to oxide-rich slag for up to 10 min as the metal color changed from reddish to bluish, and virtually all silicon and phosphorus were removed (the stage known as the clearing process). More iron oxide was added, air intake was restricted, and after about 10 min of vigorous stirring carbon oxidation began; once it was largely complete the temperature rose, large bubbles of gas began to escape, and the boiling caused red masses of slag to overflow. This high boil stage was followed by further stirring of the hot bath; then pasty masses of iron began to form at the furnace bottom, and the metal "came to nature" and was balled by a puddler into masses of 35–40 kg (30–38 cm in diameter).

Puddling was one of the most taxing labor tasks of the industrialization era: Caron (2013, 153) calls a puddler's work "virtually inhuman." Here is a description of part of the process in the memoirs of James J. Davis (1873–1947), a Welsh-born puddler in American mills who rose to be the US Secretary of Labor between 1921 and 1930:

> *Six hundred pounds was the weight of pig-iron we used to put into a single hearth ... my forge fire must be hot as a volcano. There were five bakings every day and this meant the shoveling in of nearly two tons of coal. In summer I was stripped to the waist and panting while the sweat poured down across my heaving muscles. My palms and fingers, scorched by the heat, became hardened like goat hoofs, while my skin took on a coat of tan that it will wear forever. What time I was not stoking the fire, I was stirring the charge with a long iron rabble that weighed some twenty-five pounds. Strap an Oregon boot of that weight to your arm and then do calisthenics ten hours in a room so hot it melts your eyebrows and you will know what it is like to be a puddler. But we puddlers did not complain. There is men's work to be done in this world, and we were the men to do it. We had come into a country built of wood; we should change it to a country built of steel and stone (Davis, 1922, 98–99).*

Looking back at this accomplishment, Flemings and Ragone (2009, 1964) expressed their "respect for those master puddlers who could control such a complex process with little other than their senses to guide them." Expert puddlers would process daily about 1.2 t of pig iron and produce more than one tonne of malleable iron. Eventually some of this exceptionally hard labor—prolonged manhandling of heavy iron chunks (weighing nearly 200 kg) while exposed to high heat radiating from the furnace—was displaced by mechanical arrangements, mainly by revolving furnaces first developed by the Dowlais Iron Company in England (Bell, 1884).

Iron balls taken from the furnace were squeezed to remove slag and then they went through roughing and finishing rollers (their early designs were also patented by Henry Cort) to produce iron bars roughly 2 cm thick, 6–20 cm wide, and 5–9 m long that had a mere trace (0.1%) of carbon and were cut to shorter (60–120 cm) pieces for processing into a variety of wrought-iron manufactures. Here, once more, is how Davis (1922, 86) saw it:

> *Flaming balls of woolly iron are pulled from the oven doors, flung on a two-wheeled serving tray, and rushed sputtering and flamboyant to the hungry mouth of a machine, which rolls them upon its tongue and squeezes them in its jaw like a cow mulling over her cud. The molten slag runs down red-hot from the jaws of*

> *this squeezer and makes a luminous rivulet on the floor like the water from the rubber rollers when a washer-woman wrings out the saturated clothes. Squeezed dry of its luminous lava, the white-hot sponge is drawn with tongs to the waiting rollers—whirling anvils that beat it into the shape they will.*

After 1830 an increasing share of wrought iron was turned into rails (all rails during the first 25 years of the railway era were made from wrought iron) and plates (needed for locomotive boilers as well as for the cladding for warships), and wrought iron was used in construction even by the 1880s. The Eiffel tower, a 320.75-m tall structure designed by Alexandre Gustave Eiffel (1832–1923) for the 1889 *Exposition Universelle* and completed after two years of construction in 1889, was built with wrought iron (total weight of 7,300 t) puddled at French mills and fabricated into 18,000 parts at Eiffel's factory at Levallois-Perret on the outskirts of Paris (Seitz, 2014).

Despite large increases in pig iron production steel remained a rare commodity during the eighteenth century. Blister steel was made by cementation (prolonged heating of iron bars in charcoal), and Sheffield, its principal British source, made only about 200 t/year to be used for a small range of expensive products including razors and swords. The crucible process was introduced in 1748 by Benjamin Huntsman (1704–1776), a clockmaker in search of higher-quality steel. The cementation was done in crucibles (50 cm tall and 20 cm in diameter, set at floor level) made from refractory material that could withstand temperatures of up to 1600°C. They were heated in coke-fueled furnaces and charged with blister iron and flux. After 3 h of melting the crucibles (holding about 20 kg of metal and used for three metals before discarding) were lifted from the furnace and the steel was cast to produce ingots.

Huntsman did not patent his process, and it was soon copied by other producers. Although the production of crucible steel, led by Sheffield enterprises, expanded during the latter half of the eighteenth and the first half of the nineteenth century, the resulting output was still used for relatively small items of artisanal provenience: as before, for expensive hand weapons, and increasingly for razors, cutlery, watch springs, and metal-cutting tools whose quality and dependability justified higher prices. That began to change rapidly once Bessemer introduced his method of large-scale steelmaking.

CHAPTER 3

Iron and Steel Before WW I, 1850–1914

The Age of Affordable Steel

The four key steps that raised the output and improved the efficiency of coke-fueled blast furnaces and resulted in a design that dominated for generations to come—the introduction of hot blast, capping of the furnace top, freeing of the hearth to emplace more tuyères, and redesign of typical furnace contours—were all in place by 1850. The second half of the nineteenth century saw gradual but continuing gains in performance as well as the first momentous passing of technical and market leadership from one country to another as the United States surpassed the United Kingdom in the production of pig iron and in innovative improvements in ferrous metallurgy and as Germany's iron industry developed rapidly and its output also surpassed (in 1892) the British level (Hogan, 1971; McCloskey, 1973; Paskoff, 1989; Wengenroth, 1994).

Rising demand and an improving ability to meet it with larger and more efficient operations led to the doubling of the world's pig iron output between 1850 and 1870 (to 10 Mt/year), a relative slowdown during the next three decades brought the total to nearly 30 Mt (28.3 Mt) in 1900, but by 1913 another acceleration of growth pushed the worldwide total to 79 Mt (Kelly & Matos, 2014). But, for the first time in history, this rising production of pig iron was not destined primarily either for casting of objects (ranging from pots to cannons) or for conversion to wrought iron by laborious puddling but for making steel, first in Bessemer converters and later in open-hearth furnaces.

Steel, the most desirable group of ferrous alloys traditionally available only in limited quantities and reserved for a small range of special uses, came to be produced on a large scale and at affordable costs. Iron was used by traditional societies for millennia, but only this inexpensive steel turned the metal into a ubiquitous material that could be used in much larger quantities for many old applications and that rapidly found new markets. Of these, railways and shipping were the first beneficiaries of affordable

Still the Iron Age.
DOI: http://dx.doi.org/10.1016/B978-0-12-804233-5.00003-8

steel but in a few decades steel also became an important building material, and before WW I it had also indispensable roles in the rise of new energy industries (oil and gas, electricity generation) and new modes of transportation (internal combustion engines, cars, and diesel-powered ships).

BLAST FURNACES

The most important advance of the 1850s was the introduction of regenerative brick air-heating stoves by Edward Alfred Cowper (1819–1893), whose numerous inventions and improvements also included better manufacture of candles, the writing telegraph, and the modern bicycle wheel (wire-spoke suspension with a rubber tire). His 1857 patent provided for heating of air under pressure in a brick-lined regenerator, and the heat source could be a separate fireplace or gas directed from the blast furnace (Hartman, 1980). The first stove was installed in 1859, and Cowper made design improvements for the next three decades. In 1865, when he presented his invention to the annual meeting of the British Association for the Advancement of Science, he noted that a pair of his hot-blast stoves had worked very satisfactorily, resulting in 20% higher output and higher metal quality while saving up to 250 kg of coke per tonne of iron when the blast was at 620 °C (Cowper, 1866).

Cowper's was not a primary invention but rather a rewarding application of the regenerative principle invented by Carl Wilhelm Siemens (see the next section). Regenerators were

> *enclosed in an iron casing lined with firebrick, and provided with valves to allow of the passage of ignited gas through the stove to heat it, and valves to allow the entrance of the cold blast and exit of the hot blast. The stoves are heated by the combustion of gas obtained from gas producers ... but recent experiments have been made, with the view of separating the gas from the top of the blast-furnace, from the dust it commonly contains, so that such gas may be conveniently used .... (Cowper, 1866, 177)*

Cowper found that cleaning of the blast furnace gas was not difficult and burning this CO-rich stream (containing 3.4–3.7 MJ/m^3) to produce heat in regenerative stoves provided a further boost to overall smelting efficiency. This practice soon became universal as tall columnar structures of Cowper's stoves, rivalling in height their adjacent blast furnaces, became an integral part of all iron-smelting operations. This innovation helped England's ironmasters to maintain their technical leadership as they operated the world's most advanced blast furnaces concentrated in the country's northeast (Yorkshire, Durham, Northumberland) and charged with Cleveland iron ores that were discovered in 1851 (Allen, 1981).

These furnaces combined all the latest technical advances (including caps, waste gas reuse, and vertical blowing engines), and their height and hot blast made a few of them undisputed world record holders. The tallest Cleveland furnaces were significantly taller than elsewhere: the new ones built in the late 1870s were 25.5 m tall, compared to no more than 18 m in other parts of the United Kingdom and less than 20 m in Germany, and Lucy, America's tallest blast furnace built in 1872, reached 22.5 m. The Cleveland furnace, relying on regenerative brick stoves, operated with exceptionally high blast temperatures of at least 540 °C to as much as 760 °C. During the late 1870s and the early 1880s the technical leadership passed to Pennsylvanian smelters led by Carnegie's Edgar Thomson Works (established in 1872 in North Braddock on the Monongahela southeast of Pittsburgh) whose hearths were more than 50% larger than those of typical British blast furnaces and whose operating pressures were twice as high.

Lucy, blown-in in 1878, had internal volume of 431 m^3 and daily output just over 100 t; the Edgar Thomson A furnace (originally a charcoal-fueled furnace from Michigan relocated in 1879 to Pennsylvania) was smaller (just 180 m^3), but it operated with the record blast rate of 420 m^3 a minute and used less than 1 tonne of coke per tonne of hot metal (King, 1948). In 1882 Edgar Thomson D was 24 m tall, its stack:height ratio was 0.38, and its volume surpassed 600 m^3; during the 1880s blast rates were commonly close to or above 800 m^3 a minute, and daily output of the Edgar Thomson F surpassed 300 t by 1889 while coke consumption fell to less than 800 kg/t. Before WW I America's largest furnaces continued to grow higher while their contours began to approach fairly closely the shape of a cylinder (Boylston, 1936; Fig. 3.1). The largest prewar furnace, South Works No. 9 blown-in in 1909, was 29.8 m tall with a 12.4 m stack (stack:height ratio of 0.41) but the diameter of its bosh was only 29% larger than that of its hearth.

But thanks to America's resource abundance—first, plenty of wood for charcoaling and then plenty of Pennsylvania anthracite—its ironmakers lagged Britain in fueling their furnaces with coke. Until the late 1850s (decades after British smelting was converted to coke) they relied primarily on charcoal, but the introduction of hot blast allowed the use of anthracite whose best Pennsylvanian kinds were nearly pure carbon. By the early 1860s 60% of US iron was smelted in anthracite-fueled furnaces, and although their share began to decline as iron production moved westward out of Pennsylvania, it was still just over 40% in 1880 and then decreased

Figure 3.1 New large American blast furnace from the cover of *Scientific American* of March 8, 1902.

to 12% by 1900 (Hogan, 1971). Coke has been dominant since 1875, but before 1900 about 95% of its production had been done in closed beehive ovens, circular (up to 4 m diameter) domed (height of 2.1 m) brick structures. They discharged distillation and flue gases through a central chimney, and the heat required for pyrolysis was supplied by partial combustion of coal, an inefficient process that wasted about 45% of the charge fuel.

By-product coking ovens owe their name to the capture of gases released during the coking process. They were introduced first in Europe

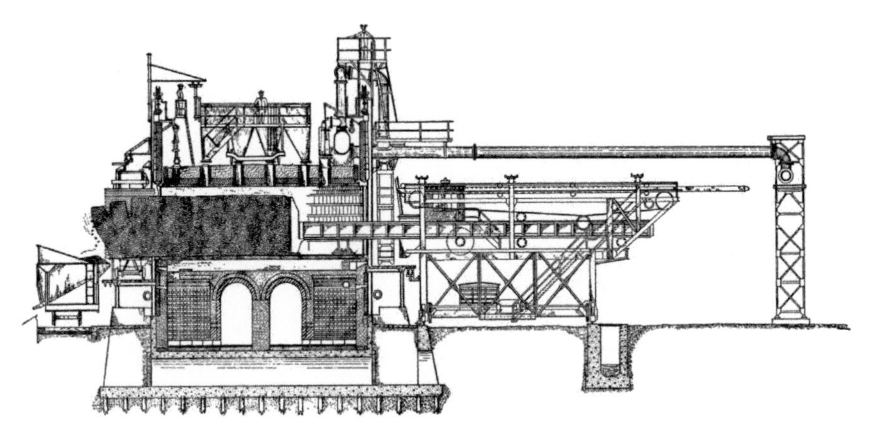

Figure 3.2 Section through an early-twentieth-century Semet-Solvay by-product coke oven showing a pusher machine loading hot coke into a car. *VS archive.*

where Carlos Otto and Albert Hüssener offered their design in 1881, refuting the prevailing opinion that by-product recovery would produce coke of inferior quality (Hoffmann, 1953). Just a year later Otto further improved the design by introducing Gustav Hoffmann's preheating of air with exhaust gases, and since 1884 Otto-Hoffmann regenerative by-product ovens lined with silica bricks, where chemicals and energy in waste gases are recovered while coke yields are increased, became the mainstay of modern coking (Porter, 1924). Semet-Solvay, Koppers, and Kuroda became the leading commercial producers of these ovens (Fig. 3.2). Initial dimensions of these prismatic chambers were about half a meter wide, about 2 m tall, and up to 10 m long, with up to several hundred of them arranged in batteries surrounded by heating and gas pipes and sealed from the air.

Their coke yield (as the share of charged coal) is higher than in beehive ovens (commonly 10–15%), and they work with a variety of bituminous coals. After the completion of the coking process (lasting roughly 1 day) the red hot fuel is pushed out mechanically and transported to a quenching tower to be cooled with water. Recovered products have a variety of industrial uses: tar, ammonia, benzol, and toluol in chemical processes and CO-rich gases as fuel and ammonia (in the form of ammonium sulfate, $(NH_4)_2SO_4$) also as a fertilizer. America's first by-product coking ovens were installed only in 1895, more than a decade after their European debut, at Cambria Steel Company in Johnstown, PA, and by 1900 there were only three other blast furnace plants with by-product coking (Gold et al., 1984). Subsequent adoption of by-product coking by the US ironmakers was rather slow: it accounted for 30% of all coke just before WW I and for more than 50% by the end of WW I, and even by 1960 the country still had more than 40 plants with some 7500 beehive ovens, accounting for 5% of total coking capacity.

In some countries conversion to coke was very gradual and charcoal use persevered for decades. In Europe, Sweden was the only major iron producer that did not switch to coke; by 1850 a quarter of the country's wood harvest was turned to charcoal (Arpi, 1953). By 1900 charcoal remained dominant but as the efficiency of smelting improved the best Swedish furnaces needed as little as 0.77 kg of the fuel per kg of hot metal, only about half of the contemporary average (Greenwood, 1907). When Japan built its first modern blast furnace in 1881 it, too, was fueled, much as the traditional *tatara*, with charcoal.

Reliable output statistics show how rapidly the American output pulled ahead of the British production after the late 1880s. In 1850 America produced 572,000 t of pig iron (ten times the total in 1810), 834,000 t before the Civil War, and, after rapidly expanding during the 1870, 4.3 Mt in 1880 (Hogan, 1971). In that year the British output was still twice as large (8.7 Mt), but then it stagnated during the 1880s while US production kept on rising. By 1889 it was 91% of the UK level and in 1890, after growing 21% in 1 year, it pulled 16% ahead of the British output (10.3 vs. 8.9 Mt), and kept this primacy for the next 80 years before it was surpassed by Soviet production in 1971.

America's primacy came from the confluence of all necessary production factors. First, plenty of wood was available for charcoaling, then Pennsylvania's metallurgical anthracite was readily available for direct use in blast furnaces as well as high-quality bituminous coal (in half a dozen states) suitable for coking. Second was the abundant supply of rich iron ores from the Lake Superior region, first from the Marquette Range (mined since 1846) and then from the Menominee, Gogebic, Vermilion, and Cuyuna ranges, and (starting in 1892) from the still exploited Mesabi range in Minnesota. Third was inexpensive transport of these ores through the Great Lakes to some of the industry's principal concentrations in the Midwest and East. Fourth was a ready supply of immigrant labor (first from Britain and Germany, after 1880 mostly from Eastern Europe) willing to work in iron and steel mills. And the fifth factor was the competitive drive (some would say the willingness to expand ruthlessly) of the industry's leading entrepreneurs, including Andrew Carnegie (1835–1919) and Charles Schwab (1862–1939).

By the year 1900, with 13.2 Mt, US pig iron smelting was 54% higher than the British output, which was only about 5% ahead of German production (its total also includes Luxembourg). During the last two decades of the nineteenth century German pig iron production had more than tripled and became more than three times that of France and four times that of Russia (Campbell, 1907). German ironmaking was heavily concentrated in the Ruhr Valley where the pioneering firms of *Gute Hoffnungshütte* (GHH) and Friedrich Krupp (1787–1826) were joined in 1867 by the company established by August Thyssen (1842–1926; Wengenroth, 1994). Together with IG Farben (the maker of Zyklon B), the last two companies later became the best-known industrial symbols of German militarism as their steel output enabled Germany to launch two aggressive wars just 25 years apart. The two companies merged in 1999, and ThyssenKrupp remains Europe's leading producer of iron and steel (ThyssenKrupp, 2015).

As already noted, global pig iron production kept on expanding during the first 13 years of the twentieth century as the US output had more than doubled from 13.2 Mt in 1900 to 28.1 Mt; German output (including that of Luxembourg) rose to 19.3 Mt, nearly twice the British production of 10.7 Mt. But regardless of their aggregate output, at the beginning of the twentieth century, all top iron-producing nations shared the basic production processes and arrangements, and hence they were able to smelt the metal with only a fraction of energy inputs required earlier in the century. British historical statistics show that the combination of larger blast furnaces, higher blast temperatures, and more efficient conversion of coal to coke resulted in steadily declining energy intensity of pig iron smelting, with typical rates falling from nearly 300 GJ/t in 1800 to less than 100 GJ/t by 1850 and to 50 GJ/t by 1900 (Heal, 1975). The nineteenth century was full of technical advances and efficiency gains, but few have equaled the performance of ferrous metallurgy.

INEXPENSIVE STEEL: BESSEMER CONVERTERS AND OPEN HEARTHS

By the middle of the nineteenth century steel was a well-known alloy and one in increasing demand, but a leading English metallurgist described accurately its restricted reach, when he wrote that in 1850 "steel was known in commerce in comparatively very limited quantities; and a short time anterior to that period its use was chiefly confined to those purposes, such as engineering tools and cutlery, for which high prices could be paid

Figure 3.3 Henry Bessemer. *Corbis.*

without inconvenience to the customer" (Bell, 1884, 435–436). Wrought iron was used for all more massive applications, and its often inferior quality was attested to by the analysis of a wrought iron disc from the hull of the USS *Monitor*, the battleship launched in 1862, and the prototype of future ironclads; Boesenberg (2006) found it to be a low-carbon, high-phosphorus ferrite with nearly 5% of silicate slag and overall mediocre quality. The first process that allowed inexpensive large-scale production of steel was invented independently on two continents, but only one name has become attached to it, that of Henry Bessemer (1813–1898), English engineer and businessman (Fig. 3.3).

Bessemer steel Bessemer announced his steel-making process in August 1856 in a paper read before the British Association in Cheltenham, and it was patented in the same year (Birch, 1968). William Kelly (1811–1888), an owner of Kentucky iron works, experimented with the blowing of air through the molten iron since 1847, eventually achieved a partial success but filed his patent only once he heard about Bessemer's innovation (Hogan, 1971). In 1857 the courts affirmed his priority within the United States (US patent 17,628) but the invention did not acquire a

Figure 3.4 Bessemer's converters. *Corbis.*

hyphenated name (such as the Hall-Héroult process of aluminum smelting or Haber-Bosch synthesis of ammonia, to note just the two most famous instances). The principle of the process patented by Bessemer and Kelly—a curious reader should consult Hogan (1971) who compares the key sentences of the two patent applications to see their shared reasoning—is easily stated.

Molten pig iron is poured into a large pear-shaped converter lined with siliceous refractory material; hot iron is then blasted with cold air that is introduced through tuyères at the converter's bottom and the ensuing decarburization removes carbon and reduces the content of impurities present in pig iron (Fig. 3.4). Cold air blowing takes between 15 and 30 min, releasing flames and smoke from the converter's top; the converter is then tilted and molten steel is poured into ladles. This process converts pig iron into steel without using any additional fuel; counterintuitively, blowing cool air through the molten metal does not cool it, but the airflow oxidizes silicon and carbon in an exothermic reaction that raises the temperature of the melt. But it was soon realized that the process works as intended only when using pig iron that is nearly phosphorus-free, as was the metal made from Blaenavon iron ore that was used by Bessemer in his work; otherwise the process leaves both phosphorus and sulfur in the decarburized metal.

An obvious (but expensive and impractical) solution was to use only phosphorus-free iron ores, but two better remedies were discovered by British metallurgists. Robert Forester Mushet (1811–1891) improved the quality of Bessemer steel, and hence made the process widely adoptable, by adding small quantities of *spiegel* iron (an alloy containing about 15% manganese) to the decarbonized iron, a process he had patented in 1856. As the name implies (*der Spiegel* is "mirror" in German), this is a bright crystalline form of iron ore with about 8% Mn and 5% C, and because manganese has a high affinity for oxygen the addition partially deoxidizes the metal and removes the impurities in slag. Removal of phosphorus from pig iron was a more difficult challenge, solved only in the late 1870s by Sidney Gilchrist Thomas (1850–1885) and his cousin Percy Carlyle Gilchrist (1851–1935).

The key to their success was to line the converter with hard and durable blocks of limestone, seal the joints with a combination of burned limestone or dolomite and tar, and add lime to the charge. Reaction of the basic lining and lime with acidic phosphorus oxides removed them from the molten iron into slag (Almond, 1981). The process was patented in 1878 and 1879 and made it possible to use Bessemer converters for pig iron made from poor (high-P) ores common in continental Europe. Andrew Carnegie, the founder of America's largest nineteenth-century steel corporation and an early licensee of the process, remarked that

> *these two young men, Thomas and Gilchrist of Blaenavon, did more for Britain's greatness than all the Kings and Queens put together. Moses struck the rock and brought forth water. They struck the useless phosphoric ore and transformed it into steel ... a far greater miracle. (Blaenavon World Heritage Site, 2015)*

A use was soon found for the copious phosphoric slag produced by the process: German engineers discovered that when ground to a fine powder it makes an excellent mineral fertilizer. Because of these complications it was only in the late 1870s when the Bessemer process surpassed wrought iron puddling in the United Kingdom, but the subsequent diffusion was rapid, and by 1890 converter steel claimed 80% of British and 86% of American production (Birch, 1968; Hogan, 1971). Bessemer steel had not only conquered all old markets using wrought iron but had made it possible to produce new castings and forgings (using large steam hammers), including those for offensive and defensive weapons, guns, and armour plate; the industry developed first in the United Kingdom whose practices were soon adopted, improved, and expanded in Germany by Alfred

Krupp (1812–1887), who visited Sheffield steel works in 1839, took over his father's company, and concentrated its production first on high-quality steel for railroad uses and, after 1859, increasingly on heavy artillery (ThyssenKrupp, 2014).

Open hearths But the conquest of the steel market by the Bessemer process was short-lived. The open-hearth furnace, a better alternative, became available during the late 1860s. Its basic version was rapidly adopted during the 1880s, and it remained the world's leading steelmaking process until the 1970s. As in the case of decarburization of steel in tilting converters, two inventors brought the process to commercial maturity, and, unlike in the former case, both of their names are sometimes used in a hyphenated form (Siemens-Martin), although most of the time the technique is called simply open-hearth steelmaking (and only the latter inventor's name is usually used in France and Russia). Carl Wilhelm Siemens (1823–1883; Charles William after he got British citizenship in 1859) was one of the quartet of inventive German brothers whose work brought fundamental advances to several branches of modern engineering. His design of the open-hearth furnace (on which he collaborated with his brother Friedrich) became a success thanks to its heat efficiency (Jeans, 1884; Riedel, 1994).

The Siemens furnace is a simple, shallow, saucer-shaped, brick-lined hearth, and it was charged first with cold, and later with hot, pig iron with small additions of iron ore or steel scrap. Its distinguishing feature was the channeling of hot gases through a regenerator, a chamber filled with a honeycomb mass of bricks to absorb much of the waste heat. Once the bricks were sufficiently heated the furnace gas was diverted to another regenerator and fresh air to be used for the furnace operation was preheated in the heated chamber; when it absorbed most of its heat its flow was switched to another heated regenerator, and this alternation minimized overall heat waste. Gaseous fuel used to heat the furnace was subjected to the same process, and hence the regenerative furnace needed four chambers for alternating operations that saved up to 70% of fuel.

The first heat-saving furnace was installed in the Birmingham glassworks in 1861 when the Siemens brothers patented the process, and by 1867 Siemens' experiments showed that this high-temperature treatment (with 1600–1700 °C in the open hearth) would remove any impurities from the charged metal. But meanwhile Emile Martin (1814–1915), a French metallurgist, not only filed his patent in 1865 but also produced the first batches of high-quality steel. As a result in November 1866

Siemens and Martin agreed to share the proprietary rights and the process was first used by several British steelworks in 1869. Early open-hearth furnaces had acid hearths, and basic furnace linings were introduced in 1879 and 1880 in France (Creusot and Terrenoire plants), giving the process its generic name: regenerative/basic steelmaking.

Unlike in the Bessemer converter with its exothermic process, heat released during the open-hearth processing was insufficient for smelting, which required an additional source of energy (coke, fuel oil, blast furnace gas, or natural gas). Open-hearth blowing lasted usually about 12 h and hot metal was poured into a massive ladle and then to individual molds to make steel ingots. During the late 1880s open-hearth furnaces began to diffuse in the United States, and in 1890 Benjamin Talbot (1864–1947), a Yorkshireman working in Pennsylvania, introduced a tilting furnace, and, with alternative slag and steel tapping, the converting process could be continuous rather than proceeding in batches. Diffusion and maturation of the technique was accompanied by increasing unit capacities: by 1900 the US steel industry had furnaces of about 30 m^2, in 1914 the maximum size was nearly twice as large, by 1944 they surpassed 80 m^2, with individual heat masses rising from 40 in 1900 to 200 t during WW II (King, 1948). The largest US open open-hearth furnaces installed during the 1950s had capacities of 600 t, 550 t, and 500 t, up to 40 times the typical heat mass of Bessemer converters (Berry, Ritt, & Greissel, 1999).

Open-hearth furnaces remained dominant for much longer than the Bessemer process: US data show that Bessemer production accounted for more than half of US steelmaking between 1870 and 1907 (peaking just short of 90% around 1880), while open hearths produced more than half of all US steel for 60 years, between 1907 and 1967, rising from just 9% of the total in 1880 to 73% by 1914 and peaking only in 1959 at about 88% (Temin, 1964). Open hearths were rapidly displaced by a combination of basic oxygen furnaces and electric arc furnaces, but a great deal of steel they made—be it in New York's skyscrapers built before 1960, in the massive dams of Roosevelt's New Deal of the 1930s, or in the first stretches of the US interstates completed during Eisenhower's presidency of the 1950s—is still with us.

Basic oxygen services were introduced commercially only during the 1950s (see the fifth chapter for details), but the electric arc furnace is the only steelmaking innovation from the pre–WW I period that is still with us. Electric arc furnaces that melt scrap metal now produce about 60% of American and almost 30% of the world's steel (again, see the details in the fifth chapter). William Siemens experimented with electric arc furnaces during the 1870s and patented his designs between 1878 and 1879, and Paul Héroult (1863–1914), one of the two inventors of aluminum smelting, operated the first commercial units producing high-quality steel from recycled metal at the very beginning of the twentieth century (Toulouevski & Zinurov, 2010). Héroult's furnaces had large carbon electrodes inserted into a furnace through its roof, a configuration that is still standard in modern designs. Before WW I there were more than 100 electric arc furnaces operating in Europe (particularly in Germany) and in North America and wartime demand for steel, combined with declining prices of electricity, pushed their number to more than 1000 by 1920 (Boylston, 1936).

Steel output Availability of processes for producing inexpensive steel from pig iron resulted in an increasing share of the primary metal being converted to steel. Right after the end of the Civil War only slightly more than 1% of US pig iron production was converted to steel, by 1880 the share was almost 30%, a decade later it reached nearly 50%, by 1900 it stood at 74%, and the conversion reached 100% by 1906 (Hogan, 1971; Kelly & Matos, 2014). And because of the increasing recycling of scrap metal (first on open-hearth furnaces, then in electric arc furnaces) American steel production began to surpass the country's pig iron output: by 1913 the former was about 12% higher than the latter. Similar shifts took place in all other major steel-producing countries, in the United Kingdom, Germany, France, and Russia.

Worldwide production of pig iron had sextupled between 1850 and 1900, from 5 Mt to more than 30 Mt, but steel output rose two orders of magnitude, from a few hundred thousand tonnes in 1850 to 28 Mt by 1900.

While historical statistics of some basic economic inputs are only approximate, post-1870 data on steel, on the global level and for all major producers, are quite reliable and portray the relentless pace of pre–WW I expansion. Worldwide steel output rose from just half a million tonnes in 1870 to 28 Mt by 1900 and to about 76 Mt by 1913, averaging an annual compound growth rate of nearly 12%. Between the end of the Civil War and 1913, US steel production increased more than 1500 times, from only about 20,000 t to nearly 31 Mt (annual growth rate of almost 16%). British steel output went from 7.2 Mt in 1875 to 7.8 Mt in 1913 (nearly an 11-fold increase), and the respective totals for Germany were about 0.5 and 17.8 Mt and for Russia 0.4 and 4.2 Mt, while Japan's 1913 steel production remained below 50,000 t.

US steel production surpassed the British total in 1887, three years before it also became the world's largest pig iron producer, and two years before it began to lead the world in iron ore extraction. By 1900 the United States was producing a third of the world's steel, that share 50% higher than the one for Germany and double the British share. America's expanding steel mills—concentrated in Pennsylvania (Pittsburgh and its vicinity), Ohio (Cleveland), Indiana (Gary), and Illinois (Chicago)—were the country's largest, well-capitalized, and highly competitive industrial enterprises. In 1869 the capitalization of American steel mills averaged less than $160,000; 30 years later it had nearly tripled (Temin, 1964). And while during the 1860s British mills could produce bars and rails about 50% cheaper than the US mills, by the beginning of the twentieth century US productivity was nearly 80% higher and the United States became a vigorous exporter of metal goods to the United Kingdom (Allen, 1979).

Not surprisingly, this expansion was accompanied by many organizational changes, including the growth of well-established enterprises, rise of new steelmaking firms, and mergers of major companies to form conglomerates on unprecedented scales. In the United States, Carnegie Steel, established by Andrew Carnegie, a Scottish emigrant to the United States, in 1892, had its beginnings in steel works in Braddock, Pennsylvania, and other neighboring mills (Bowman, 1989). By the late 1880s it became the world's largest producer of pig iron, coke, and steel, and its works constituted the largest component of the US Steel established in New York on April 1, 1901, through a merger of 10 of America's largest steel companies, producing 67% of the country's and 29% of the world's steel (Apelt, 2001).

The company's first president was Charles M. Schwab, but he left in 1904 to become the first president of Bethlehem Steel Corporation, which began as Saucona Iron Company in 1857 and was renamed Bethlehem Steel in 1899 (Hessen, 1975; Fig. 3.5). Metal made by America's second largest steel company had eventually gone into such iconic structures as New York's Chrysler Building, San Francisco's Golden Gate Bridge, and the Grand Coulee and Hoover dams, as well as into WW II Liberty ships and postwar nuclear reactors, but the company ceased operating in the Lehigh Valley in 1995 and went bankrupt in 2001 in one of the most important signs of US deindustrialization (Warren, 2008).

The availability of inexpensive steel led to an emergence of an increasing variety of alloys destined for specific markets. Their development started with Robert Mushet's experiments conducted after he was freed from his financial difficulties by Bessemer who paid his debt and gave him

Figure 3.5 Bethlehem Steel works in Bethlehem, PA. *Corbis.*

a generous annual allowance for the rest of his life (Osborn, 1952). Mushet released the first specialty alloy, soon to be known as Robert Mushet's Special Steel (RMS), in 1868. Parts and tools made from this alloy, produced with the addition of tungsten, did not need any quenching to be hardened, and yet they were better than the standard hardened and tempered carbon steel.

An early, and enduring, application of this alloy was to produce ball bearings, first to supply the increased demand that resulted from the rapid growth of bicycles during the 1880s, later to meet the needs of new automobile industry and toolmaking. Small additions of manganese resulted in brittle alloys, but in 1882 Robert Abbot Hadfield (1858–1940), after adding about 13% Mn to steel, produced a hard, wear-resistant, nonmagnetic alloy suitable for tools and bearings. Tool steel (and permanent magnets) were made by alloying with 3–4% of molybdenum, crankshaft and armor plating steel by adding 2–3% of nickel, springs by alloying with 0.8–2% of silicon, and deep-hardening steels by adding small amounts of niobium and vanadium (Bryson, 2005).

Another important alloy, patented by an American metallurgist Albert Marsh (1877–1944) in 1905 and commercially available since 1909, contained no iron. It was produced by combining 80% of nickel and 20% of chromium, and it was turned into high-resistance wires for durable heating elements, first for the newly popular electric toasters (NAE, 2000). And in 1912 Harry Brearley (1871–1948), a Sheffield metallurgist, succeeded in producing stainless steel containing 12.7% chromium and 0.24% carbon (Cobb, 2010). At the same time US metallurgists developed non-hardenable ferritic stainless steels containing 14–16% Cr and up to 0.15% C. Brearley's alloy had initially met with only lukewarm acceptance, and it was only in 1914 when a Sheffield company used it to make the first small set of cheese knives. The metal is now ubiquitous in kitchenware, food handling, engine and machinery parts, and architecture. The steel has excellent corrosion resistance but is not actually stainless; it does corrode, and Ryan et al. (2002) explained how the process usually begins at sites in the metal where there are sulfide impurities.

NEW MARKETS FOR STEEL

All of these technical advances made the four decades preceding WW I the first period in history when steel became affordable and readily available and when it could substitute for a multitude of more expensive or less durable wrought iron or wooden products and could be used on a large scale in new applications ranging from farm implements to ocean-going ships and from rails to skyscrapers. As is often the case with material substitutions, the process took a while. Perhaps the most notable example of this fact is the one that was already noted in the preceding chapter: the Eiffel Tower, one of the world's most iconic structures completed in 1889 (nearly three decades after the first commercialization of Bessemer steel), was built with wrought iron.

Before I turn to steel's use in the largest pre–WW I markets—in transportation (railways and ships) and construction (buildings, bridges)—I will survey several categories of its other important and expanding uses. Obviously, a key new market for steel was the iron and steel industry itself, engaged in unprecedented efforts to extract more iron ore, to reduce the oxides in new and larger blast furnaces, and to turn cast iron into steel and semi-finished products in new mills. Similarly, the new age of machines required a large-scale expansion of machines, tools, and replacement parts used for penetrative (drilling, boring, punching), shape-forming (turning, milling, planing), and finishing (grinding, polishing) operations (Smil, 2005). Performance of many of these tools was limited as long as they were made from plain carbon steel and Mushet's self-hardening steel.

A key advance in tool quality resulted from the experiments conducted since 1898 by Frederick Winslow Taylor (1856–1915) and J. Maunsel White at Bethlehem Steel (Taylor, 1907). Their heat treatment of tools doubled or even tripled the previous metal-cutting speeds, and, starting in 1903, these new high-speed steels allowed Charles Norton (1851–1942) and James Heald (1846–1931) to design new superior grinding machines. They were used for the first time on a large scale by Henry Ford (1863–1947) to produce many of the metal components for his pioneering Model T (launched in 1908), and their subsequent improved versions were indispensable for expanding the post–WW I car and airplane industries.

Steel in agriculture One of the earliest impacts of cheap steel, and one with momentous economic and nation-building consequences, was the metal's use in agriculture, especially the mass production of moldboard steel ploughs with soft-center steel (Fig. 3.6). Moldboard plow is an ancient Chinese invention that became common in Europe only during the early Middle Ages. The process for commercial moldboard production was patented by John Lane Jr. (1824–1897) in 1868, and it rapidly displaced the previous method of fashioning metal plows from cast iron or from old saws, the method introduced by Lane's father (John Sr., 1792–1857) in 1833 and by John Deere (1804–1886), an American blacksmith and the founder of his eponymous company, in 1837. Plows made of the three layers of steel (hard outer layer, soft inside) were better at absorbing shocks when cutting through heavy and stony soils.

Diffusion of steel moldboards was greatly helped by the adoption of sulky plows (riding plows with a steel seat) and gang plows (with two or more moldboards) starting during the late 1860s. Mechanical reapers were introduced in the 1830s—by Obed Hussey (1790–1860) in 1833 and by Robert McCormick (1780–1846) in 1837—but their mass diffusion came only after 1850 (Fig. 3.6). Without these implements it would have been impossible to open North America's central grasslands (US Great Plains and Canada's Prairies) for grain cultivation, and to harvest large fields in a timely manner. American agriculture—thanks to its decades-long march westward that created new homesteads, farms, and enclosed pastures—was also the leading consumer of barbed wire made from galvanized steel. Joseph Glidden (1813–1906), an Illinois farmer, introduced the first (and

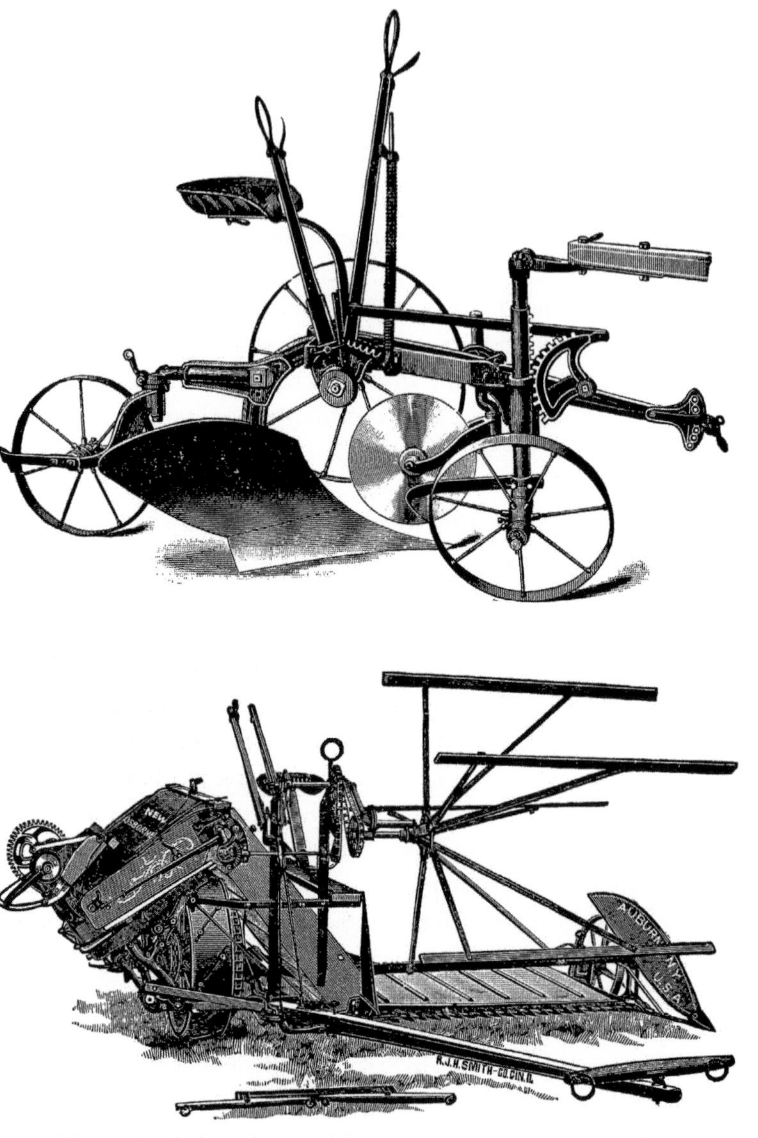

Figure 3.6 Three-wheeled steel riding plow and grain harvester (mechanical reaper) made, respectively, by Deere & Company in Moline, Illinois and D.M. Osborne & Company in Auburn, New York, during the 1880s. *VS archive.*

still frequently made) twist design in 1874, and it was followed by hundreds of permutations (McCallum & Frances, 1965).

Grain and hay production further benefited from the expanding use of other steel implements and machines, including harrows, diskers, seeders, hay balers, and horse-drawn, and since 1911 also self-propelled, combines. In aggregate, advancing mechanization, first of North American and later also of Western European agriculture, made the newly redundant labor available for expanding factory employment, and it was a key factor behind rapid urbanization and industrialization, a perfect example of positive-feedback dynamics as more steel for the rural economy led to a higher steel demand in more populous urban settings.

Traditional industries whose expansion and modernization depended on new steel components and machines included textile production (carding, spinning, weaving), paper-making, and glass-making. In 1904 Michael Joseph Owens (1859–1923) introduced the first fully automatic glass-blowing machine able to make 2500 milk or beer bottles per hour. Steel also went into new machine designs for food processing (canning), brewing, and wine-making and many other operations handling liquids and doughs that took advantage of steel tanks and containers. Canning was initially done only in glass, the first metal containers (tin-plated) were produced in the second decade of the nineteenth century, but food in steel cans became common only after 1850.

Steel in energy industries The rapid growth of fossil fuel production created entirely new markets for steel and iron. The energy sector's new demands for steel were growing especially fast in the United States. This growth is best appreciated by looking at the expansion of the country's coal and oil production: the former rose nearly sevenfold between 1850 and 1913, the latter (began in 1859) increased twelvefold between 1870 and 1913 (USBC, 1975). Coal mining required drilling, hoisting, and transportation equipment; oil and gas extraction needed drilling rigs and pipes, storage tanks, pipelines, refineries, and gas-processing facilities.

Pipeline construction was made safer thanks to the invention of seamless steel pipes by Reinhard (1856–1922) and Max Mannesmann (1857–1915) in 1885 (Salzgitter Mannesmannröhren-Werke, 2015). Soon afterward came the pilger rolling (reducing the pipe's diameter and wall thickness while increasing its length) and the combination of the two techniques has been the dominant way of pipe production ever since. And before pipelines became ubiquitous the most commonly used container holding oil was a steel barrel chosen in 1872 by the US Bureau of the Census as the standard measure of crude oil output and trade (it contains 42 US gallons, nearly 160 L). Only a small fraction of global oil trade, carried overwhelmingly by pipelines and tankers, now moves in barrels, but the anachronistic unit is still preferred by the oil industry as a measure of its output.

Another large and entirely new market was created by generation, transmission, and conversion of electricity. In 1882 Thomas Edison (1847–1931) built the first coal-fired power plant in Manhattan, and in the same year America's first small hydroelectricity station was built in Wisconsin (Smil, 2005). Steel was required for increasingly larger boilers (whose walls were lined with coiled tubing containing circulating water), steam turbo-generators (commercialized for the first time during the late 1880s and now still the world's most powerful prime movers with the largest unit capacities surpassing 1 GW), water turbines, generator halls (floors to support heavy machinery), and transmission towers and wires (modern high-voltage wires use aluminum with a steel core).

Yet another large market arose due to destructive uses of steel to create and expand modern armaments. First came the introduction of better infantry rifles, then the invention of the self-powered machine gun by Hiram Maxim (1840–1916) in 1884, and the production of steel-encased grenades and bombs and (starting in the 1880s) the construction of large and heavily armored battleships (the famous *Dreadnought* was built only in 1906). Switching to the opposite end of the scale of objects, and also indicating the breadth of new consumer items whose development took advantage of inexpensive steel, here is a brief list of small steel products invented or introduced between 1850 and 1914 that are still in demand today (Smil, 2005): paper staplers and staples (1868), shower sprinkle heads (1874; alas, too many are now in an even cheaper plastic), Swiss Army knives (1891, now in scores of designs and also in different colors), zippers (clasp lock) in 1893, spring mouse traps (1894, certainly more environmentally friendly than warfarin), and safety razor blades (1904).

And before turning to the use of steel in transportation and construction I should note that the expanded use of the metal necessitated new ways of its examination and quality testing that required many steel-based tools and devices. Steel's microcrystalline structure could be best examined under microscope (after polishing and etching the samples), and stress tests of the metal were done on machines that could exert more than 450 t in compression and 360 t in tension. But the greatest testing advances came with Wilhelm Roentgen's (1845–1923) discovery of x-rays in 1895 and with the introduction of x-ray diffraction in 1912 as these techniques allowed, for the first time, nondestructive examination of steel parts. The American Society for Testing Materials (ASTM) was established in 1898.

STEEL IN TRANSPORTATION AND CONSTRUCTION

Many important parts of traditional personal and freight transportation on land and water were made of iron, ranging (on land) from bits and shoes for horses and springs and wheel rims for coaches and wagons, to (on water) nails, brackets, belaying pins, chains, and anchors (and the all-important magnetic compass!). As important as they were functionally, in mass terms these iron components remained relatively minor accessories to the overwhelmingly wooden transportation equipment of antiquity, the Middle Ages, and of the early modern era. Only the confluence of the widespread adoption of the steam engine, the first mechanical prime mover that could power vehicles and vessels, and the availability of inexpensive Bessemer steel created new markets for the metal that soon resulted in production orders of magnitude higher than during the pre-1850 era.

The development of the steam engine for transportation became possible only after the expiry of Watt's original patent in 1800. Watt was afraid to work with high-pressure boilers, but only they could be made (relatively) light enough for mobile uses. Steam-powered vessels came first—Patrick Miller's (1731–1815) *Charlotte Dundas* in 1802 in England, Robert Fulton's (1765–1815) *Clermont* in 1807 in the United States, and the first Atlantic crossing by *Royal William* in 1833 (Fry, 1896). On land, Richard Trevithick's (1771–1833) machines on cast iron rails (in 1804) were premature, and the first scheduled railway service between Liverpool and Manchester began only in 1830. Both of these markets multiplied the steel demand for the remainder of the nineteenth century: by that time the expansion of railways in Europe and North America slowed down, but more steel was needed for ocean-going vessels (including new massive war ships), and the first decade of the twentieth century brought new demand for steel created by the world's first mass-produced car, Ford's Model T introduced in 1908.

Steel for railways Rapid expansion of European and North American railways added up to the largest market during the first decades of inexpensive steel. The railroad era began in 1830 with the world's first 56-km-long intercity link between Liverpool and Manchester. By 1900 the United Kingdom had about 30,000 km of railways, Europe had approached 250,000 km, Russia had 53,000 km, the total length of US railways had surpassed 190,000 km (including three transcontinental routes), and the global aggregate was 775,000 km (Williams, 2006). All rails laid down before the late 1850s were made from low-carbon wrought iron produced by puddling, and a significant amount of this metal for the US rails was imported from the United Kingdom even after the Civil War (and, in turn, the United Kingdom still imported Swedish wrought iron during the 1850s), and US production of wrought iron rails peaked only in 1872 (Birch, 1968).

These rails did not oxidize but they could not withstand the shocks of moving trains for a long time and had to be replaced as early as every 6–12, and later every 36, months, while steel rails would last a decade. Axle breakages were common, with the one in 1842 causing France's first great railroad disaster at Meudon where 55 passengers were burned alive (Caron, 2013). Faulty tracks on the early railways were one of the major reasons for the development and institutionalization of metal testing and standardization. Bessemer steel improved the quality of tracks as rails became its first market to conquer. The first steel rails were laid in England in 1861, and by the mid-1870s they dominated the British market, were exported to the United States, and allowed the unprecedented rail expansion on three continents (Europe, Asia and North America). In the United States iron rails still accounted for about 70% of all tracks in 1880, but two decades later their share was below 8%, and open-hearth steel was dominant everywhere.

Early rails weighed 20–25 kg/m, by 1880 the mean was about 30 kg/m and maxima reached 45–50 kg/m during the 1890s (Hogan, 1971). Using a conservative mean of 30 kg/m—to get an estimate of the right order of magnitude—would imply that between 1850 and 1900 the construction of railways consumed at least 20 Mt (and perhaps as much as 25 Mt) of iron and steel for the original tracks (Smil, 2013), and at least twice, and perhaps up to three times, as much metal for their replacement after they were in service for no more than 10 years (Ransom, 1989). For comparison, the nineteenth-century railway expansion also consumed at least 160 Mt of sawn wood for sleepers (this total included the initial emplacement and periodic track renewal) and I have also estimated—using a conservative assumption of at least 2000 t of ballast (crushed stones packed underneath and around ties) per kilometer—that some 2 Gt of coarse gravel were needed to support the tracks (Smil, 2013).

Steel also became an increasingly important component of tunnel construction, and (together with cast iron) it was an indispensable material for every other key component of expanding rail transportation, ranging from locomotives, coal and water tenders, and passenger and freight cars to switching and signaling equipment, wires, fencing, walkways, elevated

Figure 3.7 American steam locomotive: Central Pacific's model 229, first built in Sacramento in 1882. *VS archive.*

crossing, and large railway terminals (Fig. 3.7). In nineteenth-century Europe the new terminals were typically lofty, open-ended structures covered with glassed roofs, a new style of architecture so memorably depicted in a series of famous 1877 impressionistic paintings of Gare Saint-Lazare by Claude Monet (1840–1926).

Steel in shipping Shipping was the other major transportation market for steel. Small iron barges were the first metal-hull vessels built during the second decade of the nineteenth century. In 1833 the Lloyd's Register approved *Sirius*, the first iron-hulled steamer; in 1832 Macgregor Laird (1808–1861) designed *Alburkah*, an iron paddle steamer that he took on a trip to West Africa; and in 1840 Isambard Kingdom Brunel's (1806–1859) *Great Britain* was the first iron vessel to cross the Atlantic (Dumpleton & Miller, 1974). The age of steel ships began slowly during the 1860s (helped by the introduction of screw propellers in 1862) and they became common during the 1870s, even before 1877 when Lloyd's Register of Shipping accepted the metal as an insurable material for ship construction. Concord Line's *Servia*, launched in 1881, was the first large trans-Atlantic liner built of steel rather than iron, but it was in service only until 1902 (Babcock, 1931).

The demise of wooden ships took place within a single generation (no sizeable wooden vessels were built by 1900), and the race was on to build ever larger passenger steamers. By 1900 British, German, and French shipyards were building passenger ships that incorporated more than 10,000 t of steel. These efforts peaked with the construction of the largest and the fastest pre–WW I steamships, *Lusitania* and *Mauritania*, by the Cunard Line in 1907 (both 31,000 t of gross weight), and the White Star Line's *Olympic*, *Titanic*, and *Britannic*; each one of those three had gross weights of 46,000 t with hulls of plain carbon steel fastened with wrought-iron rivets (White Star Line, 2008). *Titanic* became the most famous of all the great liners after it sank on April 12, 1912, following its collision with an iceberg three to six times its mass (Lord, 1955). The quality of the steel used to make the ship's hull was among the factors suspected to have a role in the catastrophe. After the wreckage was found, remotely controlled vehicles brought up numerous artifacts as well as pieces of steel from the hull that was subject to chemical, micrographic, and mechanical analyses.

Felkins, Leighly, and Jankovic (1998) found that the metal used in the *Titanic*'s construction was probably the best plain carbon steel available in 1911 but nearly a century later would be unacceptable for any construction projects, particularly not for an ocean liner. The steel was made in acid-lined, rather than basic, open-hearth furnaces in Glasgow, and it had relatively high levels of phosphorus and sulfur (its manganese:sulfur ratio was very low, less than half that in modern steel), which tended to embrittle the metal at low temperatures and hence make it unsuitable for service in cold water (at the time of the collision the ocean near Newfoundland was only –2 °C). While it is impossible to say how much damage would have been suffered if the ship were built with the best modern steel it is most likely that, without the collision, the *Titanic* could have served, as the *Olympic* had done, for more than two decades.

Steel in the automobile industry An entirely new transportation sector was created by the introduction of automobiles during the 1880s. Designs of the earliest German models by Carl Benz (1844–1929), Wilhelm Maybach (1846–1929), and Gottlieb Daimler (1834–1900) looked like horseless carriages, with wooden bodies and steel used sparingly for a few key parts (Smil, 2005). Automobiles remained expensive because their production was entirely artisanal (small hand-made series) during the 1890s. In 1901 Mercedes 35-hp, generally considered the first modern car, had its front-mounted four-cylinder engine bolted to a pressed-steel frame, but it, too, was an expensive artisanal product made for rich buyers (Daimler, 2015). Large-scale production of affordable cars began only with Ford's famous Model T in 1908, the first car made of vanadium steel (Fig. 3.8).

Properties of this alloy were first identified and studied by J. Kent Smith starting in 1900, and after he became a consultant for Ford's company the automaker employed United Steel Company in Canton, OH, to

Figure 3.8 **Ford's Model T introduced in 1908 and discontinued in 1927.** ***VS archive.***

produce the first batch of the metal in 1906, and the steel, affordable and easy to machine, eventually constituted half of the mass of the Model T (Misa, 1995). A key part of Ford's advertisement was to stress the quality of the metal's superiority:

> *The Model T is built entirely of the best materials obtainable. No car at $5000 has higher grade, for none better can be bought. Heat treated Ford Vanadium steel throughout; in axles, shafts, springs, gears—in fact a vanadium steel car—is one evidence of superiority.*
>
> *Nobody disputes that Vanadium steel is the finest automobile steel obtainable ...We defy any man to break a Ford Vanadium steel shaft or spring or axle with any test less than 50% more rigid than would be required to put any other steel in the junk pile .... (Ford, 1909)*

As a result, the steel market for the US auto industry expanded from just a few thousand tonnes in 1900 to more than 70,000 t in 1910 (Hogan, 1971). The auto industry's demand for quality steel also led to the first efforts to standardize the alloys and to specify their permissible composition. In 1912 the Society of Automotive Engineers (SAE) chose descriptive numbers to identify 15 specific classes of steels (e.g., 10- for carbon steels, 23- for nickel steels, 61- for chromium-vanadium steels), and suffixes to indicate the share of carbon (hence 10–60 was plain carbon steel with 0.6% C, 61–25 contained 0.25% C, 0.9% Cr, and 0.18% V). In addition, the SAE also listed two key physical attributes, steel's elastic limits and

Figure 3.9 Firth of Forth cantilevered steel bridge. *Corbis.*

toughness, that were expressed, respectively, as percentages of elongation and of reduction in cross-sectional area (Misa, 1995).

Steel in construction The first, and highly visible, use of steel in construction was for new bridges, at first for relatively short crossings, later for unprecedented spans. Shorter bridges used prefabricated iron trellis girders assembled with rivets, and new suspension bridges began to use steel cables instead of wrought-iron chains (Cossons & Trinder, 1979). Perhaps the most iconic pioneering long-span bridge crossed the Firth of Forth in Scotland by a cantilevered 2529-m-long structure designed by John Fowler (1817–1898) and Benjamin Baker (1840–1907), built between 1883 and 1890 and used by the North British Railway (Fig. 3.9). The bridge required 55,000 t of the metal, most of it for its two 104-m-tall towers and massive cantilevered arms (Forth Bridges, 2013), and it has been both admired and derided.

In the United States cast and wrought iron were used for shorter bridges since the 1840s and the use of Bessemer steel began during the late 1870s. In 1879 a 270-m-long Glasgow bridge spanned the Missouri river. The famous Brooklyn Bridge was to be built with wrought iron but as inexpensive steel became available the design was changed, and the structure (its construction lasted from 1869 to 1883) has steel cables (19 strands per cable, 300 steel wires per strand, total mass of 3500 t) and steel floor beams (mass of 5000 t). In total 9500 t of steel were used to build the bridge whose main span across the East River is nearly 480 m (Feuerstein, 1998).

The use of iron in building construction predates the use of steel: cast iron (excellent in compression) formed supporting columns, and wrought iron beams were used in some smaller buildings during the 1850s, but only the availability of inexpensive high-tensile steel made it possible to build a new class of structures with the load-carrying frame of steel columns and beams and with no load-bearing walls. Steel skeletons did away with the need for massive foundations to support load-bearing masonry walls and allowed the construction of ever higher skyscrapers, starting with (by today's standards) modestly high buildings in Chicago and New York during the 1880s that used riveted I beams for columns and beams. William Le Baron Jenney (1832–1907) began the race in 1885 with his 10-story (42 m tall) Home Insurance Building in Chicago (demolished in 1931 and replaced by the Field Building). This was the first structure with a load-carrying frame of steel columns and beams (Turak, 1986).

Structural steel used for the building added up to only a third of the mass of an equally tall masonry building, and its use made it possible to design larger windows and to create more floor space. But buildings beyond four to five floors would have been impractical without mechanical elevation, the solution that first came (steam-powered) in 1857 (Otis elevator used in a five-story New York department store) and that took off only during the late 1880s with the introduction of electricity-powered elevators (for the first time in 1887 in Baltimore, then the first Otis installation in 1889 in New York) and whose rapid adoption created more demand for steel wheels, cables, and cabins. By 1890 Manhattan's World Building rose to 20 stories (94 m tall), and by 1908 the Singer Building (headquarters of the Singer Manufacturing Company) approached 200 m with its 47 stories (187 m). Other steel-based machines instrumental in the rise of modern skyscraper were construction cranes.

The second generation of skyscrapers took advantage of new H beams produced by a universal mill beam whose design (enabling the beams to be rolled in a single piece directly from ingots) was patented by Henry Grey (1849–1913) in 1897 and whose first US installation came in 1907 at the Bethlehem Steel's new mill in Saucon, Pennsylvania (Hogan, 1971). The higher strength of these beams opened the way for new high-rise designs, with the pre–WW I record held by the Woolworth Building designed by Cass Gilbert (1859–1934) and completed in 1913 as 233

Broadway (Fenske, 2008). The 60-story structure is 241.4 m tall and it embodied all the essential components of a modern skyscraper: steel skeleton, concrete foundation topped by steel beams, steel bracing to minimize swaying in Manhattan's strong winds, and high-speed elevators. Woolworth remained the world's tallest building until 1930 (when it was surpassed by the Chrysler Building and by the Bank of Manhattan and in 1931 by the Empire State Building), and a century after its construction it remains the city's (since 1983 officially designated) landmark.

Inexpensive steel transformed construction in an even more ubiquitous way, by making it possible to reinforce concrete, the most common building material of modern civilization (Shaeffer, 1992). Firing of limestone and clay at high temperatures vitrifies the alumina and silica materials and produces a glassy clinker whose grinding yields cement suitable to make high-quality concrete; the process was patented by Joseph Aspdin (1778–1855), an English bricklayer, in 1824 (Shaeffer, 1992). He named the material Portland cement because its color resembled the limestone from the Isle of Portland in the English Channel. Concrete, produced by combining cement, aggregate (sand, gravel), and water, is a material that is strong under compression (excellent for columns) and weak in tension (poor for beams), but the latter disadvantage can be eliminated by reinforcing it with iron: concrete forms a solid bond with the metal and, moreover, hydraulic cement protects iron from corrosion (Wight & MacGregor, 2011).

This composite material behaves as a monolith and makes it possible to use concrete not only for beams but also for virtually any shapes, as attested to by such modern structures as Sydney's opera house with its six large shell-like sails or the sail-like Burj al-Arab hotel in Dubai with its two reinforced concrete wings. Unlike Darby's smelting with coke or Bessemer's production of steel, reinforced concrete does not have a single identifiable commercial beginning. Reinforced concrete had eventually become the most ubiquitous building material of modern civilization, but its beginnings were artisanal. Reinforcement of stone and brick structures with iron has a long, and almost universally damaging, history (Drougas, 2009). Iron clamps and ties used to strengthen large buildings—be it Athens' Parthenon (Zambas, 1992) or Madrid's Palacio Real (González et al., 2004)—corrode in place, and the resulting iron hydroxides cause a large increase in volume that exerts pressure and leads to fissures and disaggregation of the stressed stone.

Reinforced concrete The first projects with reinforced concrete date to the 1850s—William Wilkinson's wires in coffered ceilings in 1854,

François Coignet's (1814–1888) lighthouse in Port Said—and in the 1860s they were followed by a work on concrete beams containing metal netting, and in the 1870s by patents given to William Ward, Thaddeus Hyatt, and Joseph Monier (1823–1906), a Parisian gardener whose effort began with designing bigger planters (Newby, 2001). Projects incorporating these innovations began to appear during the 1880s when Adolf Gustav Wayss (1851–1917) brought Monier's reinforcements to Germany and Austria (in 1885), and when Ernest Ransome (1852–1917) in the United States and François Hennebique (1842–1921) in France patented their reinforced designs for industrial buildings. Ransome's 1903 15-story Ingalls Building in Cincinnati became the world's first skyscraper built with reinforced concrete.

Prestressing—stretching the reinforcing bars while the freshly poured concrete (usually in precast form) is wet and releasing the tension when the reinforced concrete hardens—was first patented by P.H. Jackson in California in 1872 and in Germany by Carl Doehring in 1888. But early attempts to make it work failed due to the loss of low prestress caused by shrinkage and creep of concrete, and these problems were eventually (by the late 1920s) overcome thanks to the innovations of Eugène Freyssinet (1879–1962), a French builder of concrete bridges. The advantages of prestressing are obvious: it puts the reinforced pieces into compression, saving both materials (70% less steel and up to 40% less concrete for the same loading capacity as a segment that was simply reinforced) and making it possible to build shell-like structures.

In 1891 Chicago's Monadnock Building was the first high-rise building built with reinforced concrete, and before the century's end the quality of cement improved, and its price declined, due to the adoption of modern rotary kilns that sinter limestone and aluminosilicate clays at high temperature. Reinforced concrete became a common material for piles, foundations, chimneys, and industrial structures. In a surprisingly little-known effort, Thomas A. Edison (1847–1931), who in 1889 established his Edison Portland Cement Company in New Village, spent a great deal of time and a substantial amount of money to perfect a system of cast-iron molds that would make it possible for a contractor to pour a concrete house in a day.

The patent application was filed in 1908 and Edison advertised that his concrete houses will eventually cost just $1,200, compared to $4,000 for an average house of those years. Later a few of these cast-in-place concrete houses were built in New Jersey, Pennsylvania, Virginia, and Ohio (Steiger,

1999). Concrete was used to cast not only walls, floors, roof, and stairways but also baths, laundry tubs, fireplaces, basements resembling grottoes, and even picture frames. A few of these houses—including the prototype at 303 North Mountain Avenue in Montclair (New Jersey) still stand, massive testaments to a failed experiment (Beckett, 2012).

In contrast, Parisian apartment designs by Auguste Perret (1874–1954) and bridge designs by Robert Maillart (1872–1940) showed that reinforced concrete structures can be elegant and have a lasting appeal. In particular, Maillart's Swiss bridges—the first one built in 1901 in Zuoz across the Inn, the second one, the three-hinged arch bridge across the Rhine at Tavanasa, in 1905—have become classic examples of structural elegance (Maillart, 1935). I will close this chapter on pre–WW I iron and steel by noting another unusual application of reinforced concrete in construction: its use in cubist buildings, most notably in Josef Gočár's House (1880–1945) of the Black Madonna in Prague's Old Town designed in 1911 (Bulasová et al., 2014).

And before I begin to describe the technical innovations of the past 100 years I should note an important managerial advance of the pre–1914 period that had its origins in the American steel industry. The optimization of labor use and a quest for higher productivity began during the fall of 1880 at the Midvale Steel Company in Pennsylvania where Frederick Winslow Taylor, a young metallurgist from an affluent Philadelphia Quaker family, embarked on more than a quarter century of experiments whose original goal was to quantify all key variables in steel cutting. Taylor eventually reduced these studies to a set of calculations yielding an optimal path for a given cutting task and generalized his findings in his famous book on *The Principles of Scientific Management* (Taylor, 1911).

Taylor's quest for optimized production was often criticized as yet another tool of labor exploitation, but a careful reading of his conclusions absolves him: he was against setting any excessive quota. He reminded managers that their combined knowledge is inferior to that of the workmen they supervise and he called for their intimate cooperation with workers. Ironically, Taylor's application of his rules led to his firing from Bethlehem Steel in 1901, but his principles had eventually formed the foundation for labor optimization in modern industries worldwide.

CHAPTER 4

A Century of Advances, 1914–2014

Changing Leadership in Iron and Steel Industry

Expansion of pig iron smelting and steel production and accompanying technical advances in iron and steel industry in the 100 years since 1914 took place in seven distinct waves. The first short period was the rise of production caused by WW I demand for armaments and munitions, with new record outputs reached in Germany, the United Kingdom, and the United States. Then came the postwar fluctuations of the first half of the 1920s followed by gradual resumption of steady growth during the second half of the decade. The third wave, a downward one, was brought by the Great Depression, with pig iron output in major producing countries shrinking by as much as 75% compared to the precrisis peaks. Then yet another brief war-driven demand (1940–1945) revived the output as the United States, the USSR, Germany, and Japan reached new steel production records.

All of these periods had one thing in common: while there was some growth of unit capacities and gradual improvement in processing efficiencies, there were very few fundamental technical advances as the combination of blast furnace smelting, open-hearth furnaces, ingot casting, and ingot reheating and rolling in steel mills dominated the production. Contrary to the frequent assumption of accelerated innovation spurred by war demands, this was also true during WW I and WW II, when the rapid expansion of steel production dictated the reliance on well-proven methods that could be scaled up fast rather than on experimentation with new processes.

All of that changed during the postwar reconstruction as many advances began to transform the industry and as Germany and Japan, two countries whose iron and steel industries were severely damaged by the war, came to the forefront of technical innovation in ferrous metallurgy and processing, and as the USSR continued its mass-scale Stalinist industrialization (started in the late 1920s and set back by the war) based on heavy industrial production. This fifth period lasted until 1973, when

Still the Iron Age.
DOI: http://dx.doi.org/10.1016/B978-0-12-804233-5.00004-X

OPEC's unexpected quintupling of oil prices led to a pronounced, worldwide economic slowdown. The sixth period, the two decades between 1973 and 1993, was characterized by fluctuating or declining outputs in Europe, the USSR, and North America and by only slow growth of iron and steel production in China. As a result, the global pig iron output remained essentially constant (512 Mt in 1973, 516 Mt in 1993), and the worldwide steel output in 1993 was less than 5% above the 1973 level.

The final period has seen the truly soaring rise of Chinese steel production set against slow, or no, growth in the rest of the world, with China quadrupling its share of the global output, from 12% in 1993 to 48% in 2013 (WSA, 2014). As in so many other cases of Chinese modernization, the country's new dominance has been based on well-established techniques, and it is more notable for its stress on quantity rather than on quality. That rapid expansion carried the global steel output past the 1 Gt mark by 2004, but by 2014 it was slowing down: with iron and steel industry being in a senescent stage in the Western world and Japan and in a mature stage in China and South Korea, India is left as the only major producer with a strong growth potential, although it is highly unlikely (and, in fact, undesirable) that in the near future India will replicate China's achievements of the recent past.

Shifts in steel production during the past 100 years are an excellent indicator of changing national and company fortunes and of the uneven progress of the global economy. At the century's beginning, on April 1, 1901, Charles M. Schwab, J. Pierpont Morgan (1813–1837), and Andrew Carnegie formed the US Steel Corporation, the world's first company to be capitalized at more than 1 billion dollars (Apelt, 2001). The company produced nearly 30% of the world's steel, and the US steel mills contributed 36% of the global output. At the beginning of WW I the three leading steel producers were the United States, Germany, and the United Kingdom; at the beginning of WW II it was the United States, Germany, and the USSR.

In 1945, as the Japanese and German economies were in ruins, the United States accounted for more than 70% of the global steel output, but a decade later expanding Japanese and Soviet steelmaking began to reduce the American share. In 1970 four of the top 10 steel companies were Japanese and three were from the United States, but by 1975 the United States was still the second largest steelmaker (16% of the total) behind the USSR, and during the late 1970s the United States still had three of its steel companies among the world's top seven, with US Steel, Bethlehem Steel, and National Steel placed fourth to sixth. But the 1980s were the beginning of clear secular retreat as American steel companies, once the world's preeminent producers and innovation leaders, were falling further down the ladder of the largest enterprises.

By 1990, with the United States producing less than 12% of the world's steel, US Steel was alone among the world's top 10 steel companies (at number five), and in 1991 USX, its parent company, was removed from the list of Dow 30 industrial stocks in order to make way for Disney: displacing steel by the heirs of Mickey Mouse and operators of fairy tale lands offered an unavoidably symbolic confirmation of America's deep deindustrialization. In 1997 Bethlehem Steel, the industry's last holdout in Dow industrial, was replaced by Johnson & Johnson (Bagsarian, 2000). Low and (between 1995 and 2000) declining steel prices and large excess foreign steelmaking capacity (reaching about 250 Mt by 2000, or nearly twice the US annual steel consumption) accelerated the retreat, and between 1998 and the year 2000 more than half a dozen US steelmakers filed for bankruptcy (Smil, 2006).

The iron and steel industry's employment was cut by more than 70%: it was still above 500,000 in 1975 and 425,000 in 1980, but the total fell below 200,000 in 1991 and was just 151,000 in the year 2000, only some 20,000 higher than in the aluminum industry! By the century's end, America's steel production actually supplied a slightly larger share of the global total than in 1990 (12% vs. 11.6%), but there was no American company among the top 10 steelmakers. And by 2013, when American steel accounted for just 5.2% of the worldwide output, six out of the 10 largest steelmakers were in China, the country that produced 50% of the world's steel, and the two largest US companies, US Steel and Nucor, had ranked, respectively, thirteenth and fourteenth.

Nucor has become America's most diversified steel producer—supplying everything from highway products, building systems, joists, and decking to rebar, wires, plates, sheets, and pilings—as well as the market leader in nearly a dozen product categories (Nucor, 2014). The company operates in more than 200 locations throughout the country. In 2015 it employed more than 20,000 people and had sales of nearly $900,000 per employee. Another bit of relatively good news: US employment in the iron and steel industry has stabilized, and in 2015 it was about 150,000 workers (AISI, 2015). American steel companies have not been the only ones experiencing downgrading in the global ranking.

Nippon Steel Corporation was the world's largest steelmaking enterprise in 1990 and kept that place a decade later, but it is now (after its

2012 merger with Sumitomo Metal Corporation, creating NSSMC) in second place, behind ArcelorMittal, the company created in 2006 by the merger of Mittal Steel (founded in 1976 by Lakshmi N. Mittal in India) and Arcelor, a European steel giant that was, in turn, formed in 2002 by merging Arbed (Luxembourg), Acelaria (Spain), and Usinor, France's largest steel producer, which had merged with Sacilor in 1986 (ArcelorMittal, 2015). In 1990 four of the top 10 steelmakers were Japanese; in 2013 only two remained (with JFE, established by the merger of Nippon Kōkan and Kawasaki Steel in 2003, in tenth place).

Those European steelmakers which have not been bought by Mittal have fared much worse. In 1990 four companies ranked among the top 10—UsinorSacilor as number two, British Steel as number four, Italian Ilva group as number seven, and Germany's ThyssenKrupp as number eight—but none remained by 2013 when ThyssenKrupp ranked 21st, Riva (which acquired Ilva) ranked 46th (and in 2015 the Italian government decided to take over Ilva's operations), and British Steel did not make it to the top 50 (WSA, 2015). Just 150 years after the steelmaking revolution began, the country whose metallurgists, engineers, and workers launched it is producing less steel than Iran or Mexico, and only about a third of the Turkish output!

The first principles make the entire industry subject to significant economies of scale (costs per unit of output fall with increasing capacities): as all (blast, oxygen, electric arc) furnaces grow larger, their volumes obviously increase faster (x^3) than their surfaces (x^2), which means that capacity grows faster than investment and maintenance cost (Carlton & Perloff, 2005). Even so, high capital investments needed for new integrated steelmaking plants (especially for new blast furnaces) constitute a significant structural barrier to entry, a reality made more difficult by price fluctuations of steel products (Egenhofer et al., 2013). Not surprisingly, the greatest (and most unprecedented) expansion of new ironmaking capacities took place in China with the government's implicit direction and support: no private enterprises of that scale in China are truly private.

FROM WW I TO THE END OF WW II

In 1914 the global steel output was more than 25% below the 1913 level, but the wartime demand for weapons and ammunition led it to rise to a new record of 82 Mt in 1917 because of increases in European production. In 1917, when the United States entered the war, its pig iron production was actually about 2% lower than in 1916, and in 1918 it grew by only 1% (Kelly & Matos, 2014). The wartime global high was followed by large output fluctuations, with the total sinking to 58.5 Mt in 1919 and just 45.2 Mt in 1921, and rising to 72.5 Mt in 1920 and to a new record high of 93.4 Mt by 1925. The US iron industry was responsible for a large share of these ups and downs: in 1921 it produced 55% less than in the previous year, and in 1924 it shipped 22% less than in 1923.

Notable changes introduced by the US ironmakers included a further steepening of the bosh angle, increase in furnace heights and hearth diameters, and efficient removal of dust (King, 1948). In 1918 the South Works No. 6 furnace had a total volume of 776 m^3, bosh angle of just over 82°, and hearth diameter of 6.25 m; it produced about 605 t of pig iron a day (with coke:iron ratio of 0.89 by weight), its blast rate was 1250 m^3 at 600 °C, and its charge was just Mesabi iron ore. The next generation of furnaces during the late 1920s extended the hearth diameter to 7.5 m: Ohio No. 2 had a volume of 1205 m^3 and produced about 980 t/day (coke:iron ratio of 0.8); its blast rate was 2045 m^3 and Mesabi ore made up about 70% of its charge. Larger furnaces required improvements in handling and charging of raw materials and removing the hot metal in large ladles. The capacity of the first brick-lined ladles grew from nearly 10 t to nearly 70 t; during the 1920s came the ladles with capacities of 125–150 t.

Another early American contribution solved a troublesome problem of cleaning blast furnace gases to a much higher degree of purity than it was possible to do by primary washers. The first electrostatic precipitators were designed just before WW I by Frederick G. Cottrell (1877–1948) when he was working at the US Bureau of Mines, and the first commercial installation was at a manganese plant in 1919, with blast furnace application coming soon afterwards (King, 1948). Precipitators rely on a simple ingenious method of charging plates or tubes (cathodes) and cables spaced between them (anodes), with dust deposited on cathodes and regularly shaken off for removal. This efficient cleaning method began in the 1920s and was later adopted to remove fly ash from coal-fired power plants (with efficiencies up to 99.9%, replacing dark plumes by condensed CO_2-rich steam) and cement factories.

New all-time production records were finally set, after four years of steady increases, in 1929: 38.7 Mt of pig iron and 50.3 Mt steel in the United States and 98.5 Mt of pig iron and 120.8 Mt steel worldwide. Maturation and expansion of car industry was a major source of new demand for steel as the designs shifted to fully enclosed, streamlined automobiles whose bodies were made of sheet steel and whose heavier chassis added to the total mass of the metal needed per vehicle. This conversion was aided by the introduction of a "universal" steel, a chromium-molybdenum alloy: several carbon grades of this steel could be used in all parts of vehicles, lowering the cost of procurement and production (Misa, 1995). In the United States three other notable drivers of steel demand during the 1920s were the widespread construction of new skyscrapers, the doubling of electricity generation, and the massive acquisition of household appliances, including washing machines, vacuum cleaners, and refrigerators, whose sales rose from just 10,000 to nearly 800,000 units between 1920 and 1929 (Hogan, 1971).

But this rise was interrupted by the worst economic crisis of the modern era: industrial production contracted (in the United States by 46% between 1929 and 1932), and iron and steel industry, its fundamental precursor, suffered an even greater decline. In the United States the 1932 output of 12.4 Mt of steel was lower than at any time since 1904, and by 1938 it was still below the 1915 level (Kelly & Matos, 2014). Between 1931 and 1936 55 US blast furnaces, capable of producing nearly 6.5 Mt of pig iron, were abandoned, and the number of idle furnaces peaked in 1932 with 235 units (only 44 were operating). The total number of active furnaces in 1929 was not surpassed until 1939 (Hogan, 1971).

Similar output declines, furnaces abandonments, and idling were experienced in the United Kingdom and France. Worldwide output fell by nearly 60% between 1929 and 1932, but a new record was set in 1937 as the USSR, Germany, and Japan expanded their production. But even during the crisis years, car industry was laying down the foundation for future expansion through continued optimization of assembly lines (resulting in higher productivity and lower car prices) and the diffusion of American designs abroad: in 1931 Ford opened an assembly plant in the United Kingdom (in Dagenham in Essex), and in 1932 it opened one in the USSR (in Nizhny Novgorod), and during the 1930s the notable visitors who studied the organization of Ford's massive River Rouge plant included Giovanni Agnelli (1921–2003; future CEO of FIAT) and Kiichiro Toyoda (1894–1952; future president of Toyota Motor Company).

In the USSR the Stalinist quest for rapid industrialization was based on the fastest possible development of heavy industries in general, and iron and steel in particular. By 1928 Soviet iron and steel production was still below the 1913 level (3.3 vs. 4.2 Mt pig iron, 4.0 vs. 4.2 Mt steel), and the first five-year plan (1928–1932) envisaged increases to 10 Mt of iron and 10.4 Mt of steel, but only about 60% of these goals were met, with 6.2 Mt of iron and 5.9 Mt of steel (Dunayevskaya, 1942). During the 1930s the USSR established a modern steel base in Magnitogorsk (southeastern Ural region) based on the design of the US Steel mill in Gary, Indiana. By 1937 the country produced 17.7 Mt of steel, and Magnitogorsk, deep in the rear, was indispensable for producing steel for the country's victorious war against Nazi Germany.

German wartime steel production peaked at 19 Mt in 1918; the output was around 15 Mt in the mid-1920s. In 1926 German steel industry coalesced into *Vereinigte Stahlwerke*, a conglomerate of metallurgical and mining companies modeled on US Steel; it was partly nationalized in 1932, and it became the principal provider of the indispensable metal for the country's large-scale rearmament that followed Hitler's rise to power (in 1933) and then to an overt preparation for and conduct of a new European war. Steel output rose from the crisis low of 6 Mt in 1933 to the wartime peak of 22 Mt in 1940 and then declined to 18 Mt in 1944 as a result of the Allied bombing. But large aggregate pre–WW II German gains were much less impressive in relative terms: in 1937 the United States consumed nearly 400 kg/capita, Germany about 290 kg, and the USSR about 105 kg.

And even during the crisis years of the 1930s new demand for US steel was created by the railways switching from steam locomotives to diesel traction, and from heavy road transport changing from gasoline- to diesel-powered trucks. The first shift was exemplified by GM's powerful locomotive (448 kW) pulling *Pioneer Zephyr*, the world's first stainless streamlined train, which set new average intercity speed records in 1934 (Smil, 2006). US adoption of diesel trucks lagged behind Europe, and Kenworth Motor Truck Company was the first maker of American diesel trucks, starting in 1933. The United States entered WW II on December 9 after the Japanese attack on Pearl Harbor, but prior to that it was already helping to arm the United Kingdom and it had begun to mobilize for what was increasingly looking to be an unavoidable conflict.

Most of the metal required by America's war effort came from capacities built since the 1920s, but new blast furnace designs, whose output dominated the first postwar decade before the Japanese ironmakers built their new models, were introduced at two new furnaces at Edgar Thomson Works in Braddock, Pennsylvania, in 1943. Each furnace had a total volume of 1446 m^3, a hearth diameter of 8.25 m, and a hearth area of 53.5 m^2, produced 1260 t of hot metal daily using a blast volume of 2440 m^3, and was charged only with Mesabi iron ore. Steel-based performance of America's war procurement was impressive. Direct mill shipments to shipbuilding industry rose more than 12-fold between 1940 and 1943 (Hogan, 1971). By 1944 American shipyards were launching 17 times as many ships as in 1939, production of munitions was 20 times, and aircraft completion was 28 times higher (Tassava, 2008).

But the overall steel output rose much less—by 28% in 1940 (to 78 Mt), and its wartime peak (81.3 Mt in 1944) was less than 5% above that level—because the metal, and all other requisite raw materials, previously used for consumer items ranging from cars to refrigerators was diverted to war uses. A single comparison illustrates steel's contribution to American victory: during WW II the United States produced 6.6 times more munitions than Germany (Gatrell & Harrison, 1993). In 1944, the peak wartime production year, the United States produced nearly 50% of the world's steel, and that share rose to nearly 72% in 1945 as the blast furnaces and steel mills in the two defeated powers, Germany and Japan, were severely damaged during the war.

AMERICA'S POSTWAR RETREAT

American designs dominated the development of new blast furnaces until the late 1950s. The first important postwar ironmaking innovation was the introduction of smelting under pressure (70–140 kPa) by the Republic Steel in 1947–1948. This innovation led to considerable savings of coke. Other operational advances of American pig iron industry of the 1950s included the use of highly beneficiated ores, enrichment of blast air by oxygen, injection of gaseous or liquid fuels into blast furnaces, better refractories, automated plug drill, carbon hearth lining, and better automated process control (Gold et al., 1984).

Obviously, America's dominance of global steelmaking could not last, and it began to decline almost immediately with the postwar recovery of Europe and, a few years later, of Japan, and with continued industrialization of the USSR. By 1955 the United States produced 39% of all steel, and by 1973 its share fell just below 20%. The United States remained the world's largest steel producer until 1970; the USSR claimed the top spot during the next 2 years, and the United States regained the top position just for 1 year in 1973 when it produced the record amount of 137 Mt. As it soon became clear, that achievement reflected the industry's large capacity and its past accomplishments; it was not the foundation of further progress. Just the opposite was true.

As all modern economies were affected by OPEC's sudden oil price rise in 1973–1974 the US pig iron retreated: by 1975 it was 21% below its 1973 peak, and, after a few years of fluctuation, it began to decline even more rapidly in 1982 (dropping by 41% in a single year). After reaching the peak output of 137 Mt in 1973, America's raw steel production was halved by 1982 (with more than 60% of that drop taking place in a single year), as was the total number of workers in iron and steel industry, to fewer than 200,000 (Haller, 2005). Stagnation and decline of America's steel production was strongly related to the retreat of domestic carmaking, the country's largest manufacturing industry, exemplified by the shares of the market taken by GM, the world's largest auto maker: the company's dominance peaked in 1960 with nearly 50% of all cars sold, by 1980 its share was just above 40%, and by the year 2000 it fell to less than 30%.

Hall (1997) called the years between 1975 and 1989 "Melting down: The end of 'Big Steel' in the United States." The decades-long pattern of an industry dominated by a few large integrated iron-and-steel companies enjoying stable market share, generally rising demand, and control over pricing disappeared with a speed that was totally unanticipated by the industrial leaders and the powerful unions alike. General obsolescence of America's ironmaking and steelmaking plants (most notably, as I will explain later in this chapter, their continued reliance on open hearths and tardiness in introducing basic oxygen furnaces), complacency of the leadership of large companies used to decades of market dominance, and high wages driven by aggressive, strike-prone unions collided with weakening domestic demand, rising low-cost high-quality imports, the emergence of highly competitive domestic mini-mills, and new requirements for more stringent environmental controls.

The rest of the twentieth century saw a few years of temporary production increases, but by the century's end the US pig iron output of 47.9 Mt was just 52% of the record 1973 level, and afterwards it continued to fall, sinking to just 19 Mt during the economic crisis of 2009. Just 2 years later it was once again above 30 Mt, but in retrospect it is painfully clear that the American industry was forced to make a belated fundamental break, moving away from traditional integrated (blast furnace-based) steelmaking and rebuilding itself as a vigorous recycler of scrap metal. Hall (1997, 336) described the new situation correctly when he concluded that

> *the term "steel industry" itself may be obsolete. The industry is, on one hand, part of a broader metals or even materials industry ... on the other hand, the industry is seeing vertical de-integration as the historic links among raw materials, steelmaking, rolling, processing and distributing become unravelled.*

American steel industry remained depressed during the 1980s, but by 1994 it was above 90 Mt/year. By the century's end the United States had 40 blast furnaces, compared to 452 in 1920, and although their productivity was an order of magnitude higher, their total pig iron output in 1999 was less than 40% above the level in 1920. But as EAF production prospered, total steel output rose to 102 Mt in the year 2000 before it resumed, once again, its decline during the first decade of the twenty-first century: during the economic crisis of 2009 it fell to just 59.4 Mt, and although the output was back to 87 Mt in 2013, it accounted for only 5.4% of the global output, slightly better than the record low of 4.8% in 2009 (Kelly & Matos, 2014). But the jobs have not returned: productivity gains through technical advances and automation made many mills more competitive and assured their continued survival, but they have also resulted in mass-scale job losses. The total employment in the industry fell from 521,000 in 1974 to 204,000 by 1990, and since the year 2000 it has been mostly just above 150,000 as the remaining integrated plants employ a fraction of their former labor force.

America's largest integrated steel mill, Gary Works in Gary, Indiana, has been in operation for more than a century, specializing in high-quality steel production (Fig. 4.1). In 1950 the mill employed 30,000 workers and it had a capacity of 6 Mt/year; in 2015 it employed just 5000 workers but its annual capacity was 7.5 Mt (USS, 2015). In 2015 the company closed its coke-making operation, laying off 300 workers. And the country's largest blast furnace, Bethlehem's old L unit at Sparrow Point plant in Maryland, was shut down by its Russian owners (Severstal) in July 2010 (Bethlehem Steel had already filed for bankruptcy in 2001), and it was demolished in January 2015 (Wood, 2015). Closures of traditional integrated (blast furnace-steelmaking) works and the rise of minimills also took place in Europe and Japan, but American job losses were relatively larger (77% decline between 1974 and 2000) than in Japan or Germany, where the cuts amounted to, respectively, 63% and 67% (Herrigel, 2010).

Inevitably, these changes also altered America's status as a steel trader. The country remained a net steel exporter until 1958. Then came a slow rise of cheaper imports to about 10% of total consumption by 1973 and to almost 24% by the century's end, while recent net steel imports (excluding semi-finished products) have remained mostly between 15% and 18% of apparent consumption (Kelly & Matos, 2014). And, of course, the great retreat also meant the loss of the long-held global primacy. Post–WW II expansion of the Soviet steel industry kept on narrowing the US lead: in

Figure 4.1 Gary Works of the US Steel Company in Gary, Indiana. *Corbis.*

1950 the USSR produced 31% of the US output; by 1960 it was 72%. After topping the United States for the first time in 1971, the country remained the world's top steel producer between 1975 and 1991 (peak output of 163 Mt in 1988), and only when the union unravelled (just before the end of 1991) and the industrial production of its successor countries collapsed did Japan rise to the top position, which it held for just 4 years before China became the world's largest producer in 1996 and has remained ever since.

JAPAN IN THE LEAD

Japan's first integrated modern iron and steel plant began its production only 48 years after the country's opening to the world, in 1901 with the blowing-in of Higashida No. 1 blast furnace at Yawata Steel Works in northern Kyūshū (Yonekura, 1994). Steel output rose during WW I, and after the stagnation of the 1920s Japan's steel demand during the 1930s was driven primarily by the conduct of war in China and by the preparation for attack elsewhere in Asia and on the United States. Japan became self-sufficient in steel production during the 1930s, by 1937 it produced 6.4 Mt, and by 1938 nearly 40% of iron and steel consumption was claimed by the military (Yasuba, 1996). Even so, there could be no better material indicator of Japanese madness in attacking the United States than the fact that the country's steel production in 1940 was less than a tenth of the American rate (6.9 vs. 78 Mt): in material terms Japan lost the war even before it began.

Despite the ban on US scrap imports (which amounted to 2.17 Mt in 1939), wartime Japan was able to keep steel production close to the pre-war level until 1943 (Emi, 2015). Its largest steel plant, Imperial Iron and Steel Works in Yawata, was the first target chosen for a mass attack on Japan's main islands, but the June 1944 raid caused almost no damage (Polmar & Allen, 2012). But this plant and other steelworks were eventually damaged before the war ended in September 1945, and by the end of 1946 the country had only three working blast furnaces, compared to more than 20 before the war. Naito, Takeda, and Matsui (2015) published a detailed account of the postwar recovery (it was surprisingly fast, with 16 blast furnaces operating by 1950) and of the following phases of Japan's rise to the world leadership in ferrous metallurgy.

Postwar development of Japanese steel industry was guided by three rationalization plans formulated by the Ministry of International Trade and Industry and extending between 1951 and 1965: they included tax breaks, depreciation allowances, tariff exemption for the industry's imports, support for exports, speedy licensing of foreign patents, and access to cheap credits (Elbaum, 2007). The two rationalization waves of the 1950s reduced coke inputs and boosted productivity. The first major postwar project, Chiba Works of Kawasaki Steel (now JFE Steel) on the northeastern shore of Tōkyō Bay, began its operation in 1953, and it was followed by construction of new large (2000 t/day) furnaces in half a dozen mills.

Japan was an early adopter of two key steelmaking innovations: in 1955 Sumitomo Metal licensed the continuous casting method, and the first basic oxygen furnaces began operating in 1957 (50-t BOF at Yawata Steel), and oxygen steelmaking surpassed open hearths by 1965 (Emi, 2015). British steel output was surpassed in 1960 and West Germany's production 3 years later, and in 1970 Yawata Iron & Steel and Fuji Iron & Steel merged to form Nippon Steel Corporation. Demand for steel was driven first by postwar reconstruction and then by Japan's rapid urbanization and construction of enviable public and intercity transportation and, to an increasing degree, by exports of steel-intensive products.

First, starting in the late 1950s, Japan became a global leader in production of cargo ships and giant oil tankers in particular, and then came the Japanese car exports, with Honda and Toyota pioneering the conquest of the US market (Smil, 2006 and 2013). Makers of industrial plant equipment (Mitsubishi, Mitsui, Ishikawajima) and heavy machinery (Komatsu, Sumitomo) were other key exporters with large demands for quality steel.

During the 1960s Japan's pig iron production rose nearly sixfold (to nearly 70 Mt), and by 1973 the country had 60 operating blast furnaces (all of them using heavy oil injection) and only the USSR (also relying on large, highly automated but less efficient furnaces) produced more pig iron. In 1973 Japan's crude steel production reached 119 Mt, equal to 16% of the global output—but to measure the importance of Japanese steelmaking merely by the country's aggregate output or by the number of years it was the world's largest producer would profoundly underestimate its contribution to the post–WW II advancement of the industry.

By the late 1950s the US ironmakers still dominated the industry in terms of output but Japan had emerged (just a few years after the country finally surpassed its prewar economic performance) as the most innovative producer of pig iron. As in so many other instances of the country's technical progress, the Japanese ironmakers first mastered the American experience and then they proceeded to build blast furnaces distinguished not only by their unprecedented size but also by their innovative features (ranging from automated material handling to computerized furnace controls and optimization of processes through operations research) and overall efficiency and productivity.

Between 1960 and 1990 every one of the successive 16 furnaces to hold the world record for internal volume was built in Japan. The world's largest blast furnace, No. 2 furnace at Nippon Steel Oita Works, with a hearth diameter of 15.6 m and volume of 5,070 m^3, was blown-in in 1976 (Fig. 4.2). Rising furnace size and advances in operation lowered the reducing agent rate (coal and oil) to 494 kg/t of hot metal, the world's lower level, and Japanese ironmakers pioneered, among other innovations, high-pressure process equipment, large hot stoves, bell-less charging arrangements, and computerized furnace controls. Expansion of improved ironmaking was accompanied by Japan's eager adoption of the new steelmaking system that combined basic oxygen and electric arc furnaces with continuous casting and improved methods of rolling (Ogawa, 2012). This resulted in a rapid reversal of productivity rankings. While in 1960 Japan's average labor requirement was almost 49 man-hours to produce a tonne

Figure 4.2 Nippon Steel's Oita blast furnaces No. 1 and No. 2 (on the left) at Oita Works in eastern Kyūshū. *Reproduced by permission from NSSMC.*

of steel, nearly three times the US mean of 16.5 h, by 1973 the Japanese mean was below 9.5 h and the US steelmakers averaged just over 11 h (USBC, 1975).

Japanese crude steel production was less than 6% of the US output in 1950, but two decades later it was equal to nearly 80% of the US level, and it surpassed it in 1980. Its production originated in large modern coastal steelworks designed to receive raw materials transported by large bulk carriers from overseas (Australian and Latin American ore, North American and Asian coal), convert them in blast furnaces whose inner working volumes were mostly in excess of $3000\,m^3$, and process pig iron in basic oxygen furnaces. Japanese mills were among the earliest adopters of continuous casting and also became the world's leaders in the longevity of furnace operations and in their energy efficiency (Naito, Takeda, & Matsui, 2015).

Japanese steelmakers also led the effort to pretreat hot metal in order to reduce the presence of silicon, sulfur, and phosphorus before decarburization in converters. They also pioneered the use of process computers to supplant human control in operating blast furnaces and steelmaking furnaces, and they introduced advances in refractory formulation and installation. And as steel demand began to stagnate, some of them looked for new business opportunities: NKK diversified into the treatment of waste materials and recycling, water treatment, municipal waste incineration, air pollution control, energy conservation, clean energy production, and soil remediation.

Japan's twentieth-century steel production peaked in the same year as the US output (in 1973 at 119 Mt), and then—affected by oil price rises of the 1970s, the high value of yen after 1986, and the rise of Chinese steelmaking in the 1990s—it fluctuated mostly between 95 and 110 Mt until 2003, before reaching a new high of 120 Mt in 2007, but it retained its focus on performance, efficiency (low coke rate, pulverized coal injection), and the quality of both material inputs and finished steel products. The industry's labor productivity rose from 450 t/worker in 1975 to 750 t in 1990, to 1400 t in the year 2000, but it has been largely stagnating since that time. The number of blast furnaces fell from 40 in 1985 to 27 by 2013, but 13 of them were converted to have volumes of more than 5000 m^3, and their productivity reached a new high at 1.94 t/m^3. After Nippon Steel merged with Sumitomo Metal, the new company decided to close the No. 3 furnace at Kimitsu by the end of fiscal year 2015.

CHINESE DOMINANCE

The pace of China's ascent to the dominant place in iron and steel production has been determined by the country's enormous shifts in basic industrial policies. When the Communist Party of China took the full control of national government and established the People's Republic in October 1949, the country was producing just 158,000 t of steel, but then its economic policies copied Stalinist stress on the development of heavy industries. In 1957, after the end of the country's first five-year plan, China's steel output reached 5.3 Mt, heavily concentrated in the Northeastern provinces where Japan developed the industry during its pre-1945 occupation of Manchuria (Tang, 2010).

But, in a typical Stalinist fashion, a small steel mill established in the western suburbs of Beijing in 1919 was greatly expanded and Shougang steelworks eventually became the country's largest steel producer (with annual capacity of 10 Mt, employing 200,000 workers), and for more than half a century it kept on polluting the capital. The enterprise was closed down only before the 2008 Olympics, when the government embarked on a massive cleanup of the capital (Fig. 4.3). The company's new site,

Figure 4.3 Shougang iron and steel plant in the western suburbs of Beijing. The photo was taken in February 2012, 4 years after the plant was closed before the 2008 Olympics. *Corbis.*

with two very large (>5000 m^3) blast furnaces, is on Caofeidian, an artificial island reclaimed from Bohai Bay, east of Tianjin in Hebei provinces.

As the Sino-Soviet rift deepened during the late 1950s the Stalinist model appeared too slow for Mao Zedong (1893–1976), who wanted to catch up, even surpass, the performance of leading Western economies in a matter of years. While Nikita Khrushchev was telling the Americans "We will bury you," Mao devised a much bolder version of catching up and surpassing, a delusionary Great Leap Forward. In 1958 the Communist Party's Central Committee

> *put forward an inspiring slogan which calls on the people of the entire nation to exert their utmost so that China can surpass Britain within 15 years or in less time in output of iron, steel and other major industrial products. In other words, within that period China is to be transformed from a backward agricultural country into an advanced, socialist industrialized one. The Chinese people, filled with firm confidence and enthusiasm, are striving for the fulfilment of the Party's call.*
>
> ***(Huang, 1958, 1)***

And the next step was to surpass the US industrial production as China pursued its astonishing dash toward global supremacy: Mao, a dutiful (albeit Stalin-hating) Stalinist, was obsessed with steel production, but his ignorance of realities led him to conflate pig iron and steel and to call for production methods that were doomed to fail. In May 1958 Mao claimed that "with 11 million tons of steel next year and 17 million tons a year after the world will be shaken" (Mao, 1969, 123), but as China had just a handful of Soviet-designed modern iron and steel mills built during the first five-year plan, his solution to make the Leap work was a mass mobilization of peasant labor (Wu & Ling, 1963).

Some 20 million peasants were forced to abandon their fields and open up 110,000 small mines to produce poor-quality coal used mostly in smelting of pig iron. Tens of millions of peasants were ordered to dig local deposits of low-quality iron ore and limestone, and in order to produce enough charcoal they were forced to cut down not only scarce forest trees but also orchards and groves, intensifying the country's deforestation. These low-quality ingredients (and often also any scarce scrap metal, even pots and pans) were charged into simple clay "backyard" furnaces, traditional Chinese clay structures that were used for some two millennia for small-scale local pig iron production. Vastly exaggerated claims of the Great Leap years make all numbers suspect, but as many as 600,000 of these clay furnaces may have been built in 1958 and 1959 (Smil, 2004).

In 1959 Zhou Enlai (1898–1976) said that in the previous year those furnaces produced 4.16 Mt of usable pig iron, or 30% of that year's total output, and it was estimated that there was an additional 4–5 Mt of unusable inferior-quality metal (Wagner, 2013). Naturally, this smelting could not produce any steel—just lumpy cast iron heavily contaminated with slag and with high carbon content, and hence very brittle and unfit for making even simple farming implements—but the Chinese reports, as well as the numerous writings by foreign observers and Sinologists, keep referring to the metal as steel from backyard furnaces, an instructive indication of the lack of basic understanding of material realities of the modern world (Li & Yang, 2005).

Other mass campaigns of the Leap years diverted labor from agriculture to wasteful industrial efforts, but due to their high labor requirements backyard iron furnaces were a key reason for the sharp decline of crop harvests and the world's largest Mao-made famine, which claimed the lives of more than 30 million people (Smil, 2000). The Chinese Communist Party continues to blame poor weather as the famine's principal cause, but the official Chinese statistics show that similar or more extensive droughts and floods had only a marginal effect on the country's harvest once the agriculture was privatized during Deng Xiaoping's modernization drive of the 1980s (Smil, 2000).

Once the Great Leap ended (in 1961) backyard furnaces were abandoned, and during the 1960s steel production from older Soviet-built mills and new mills based on Soviet know-how (Sino-Soviet rift turned into undisguised hostility by the mid-1960s) grew 2.5-fold, from less than 7 to nearly 18 Mt between 1962 and 1970 (NBS, 2000). In 1973, 15 years after the call for surpassing the United Kingdom, China's modern steel mills produced 22.3 Mt compared to the British output 26.7 Mt in: what mattered is not that the output came within 17% of its 1958 goal but the enormous damage done by the Great Leap effort. Politically and socially China was in turmoil during the 1970s (end of the Cultural Revolution, Mao's death in September 1976, succession struggles), but steel output had doubled to 37 Mt.

During the first decade of Deng Xiaoping's (1904–1997) economic reforms, steel production grew by nearly 80% to reach 66 Mt in 1990 as China began to benefit from foreign expertise and designs: most importantly, the largest iron and steel mill of the early period of modernization, Shanghai Baosteel (now the world's fourth largest steel producer), received a lot of technical support from Nippon Steel. During the 1990s China's steel output, once again, almost doubled, to 128.5 Mt, but these three successive doublings or near doublings were just a prelude to China's enormous steel leap during the first decade of the twenty-first century, when the output more than quadrupled, from 151.6 Mt in 2001 to 638.7 Mt in 2010.

Nobody anticipated such a dramatic increase: a 1999 paper authored by Chinese experts forecast 330 Mt of steel in 2010 and 650 Mt by 2030 (Lo et al., 1999), but the latter total was surpassed already in 2011. Early phases of this remarkable expansion included the construction of hundreds of small furnaces (with internal volumes of just 200–500 m^3) whose relatively inefficient operation required 700 kg, or more, of coke per tonne of pig iron (Okuno, 2006). Small sinter plants (average area of less than 60 m^2 compared to 200 m^2 in the EU and nearly 350 m^2 in Japan) were also common, as were electric arc furnaces with capacities of no more than 30 t. Many large furnaces have been constructed since 2005, but by 2013 China still had 19% of blast furnaces with volumes less than 1000 m^3 (compared to the global average of 4%) and 59% of furnaces larger than 2000 m^3, compared to the global mean of 79% (VDEh, 2013).

The highly fragmented state of China's steel industry is best illustrated by the fact that by 2010 the country had about 1200 steel producers, of which only some 70 were medium- and large-sized enterprises, and all of the major ones, except for Shagang Steel, were state owned (Tang, 2010). But larger companies have been rising in the global rankings: in 1990 China had a single steel company among the top 25, in 2000 it had four, and in 2013 it had 11 (WSA, 2015). Performance of these large producers has improved with the use of high-quality imported ores and high-strength coke and with the adoption of pulverized coal injection.

By 2005 the best mills, such as Baoshan Steel in Shanghai, operated with a coke rate as low as 290 kg/t, augmented with up to 200 kg/t of injected coal dust, very much in line with the best Japanese experience. China has some 8000 iron ore mines, but most of them are small and ship low-quality hematites (Fe content of just 30–35%) with a high share of impurities, and the country's high dependence on iron ore imports has made it suddenly the dominant factor on the global market for the world's second largest (after crude oil) traded commodity. This change was rapid: in the year 2000 China bought about 15% of all traded iron ore, by 2005 that share was up to 36%, and by 2013 it was 64% (WSA, 2014; Fig. 4.4).

Since 2010 the growth of China's steel output has continued (albeit at a much reduced pace), with nearly 780 Mt produced in 2013 and just over

Figure 4.4 Imported ore in Qingdao port (Shandong province) in China. *Corbis.*

800 Mt in 2014, when China's output accounted for 49% of the world's steel. Unlike in other major steel-producing countries, all but a small fraction of China's steel output comes from integrated enterprises, with electric steel contributing less than 10% of the total (8.8% in 2013). Of course, China's infrastructural expansion has demanded unprecedented amounts of raw materials: nothing illustrates this demand better than the fact that in just 3 years China emplaced more concrete in its buildings, dams, and transportation links than the United States did during the entire twentieth century (Smil, 2013).

Not surprisingly, the single largest finished products of China's steel industry have been reinforcing bars (Fig. 4.5). By far the largest accumulations of reinforced concrete are in massive gravity dams on the world's major rivers: the record goes to China's Sanxia (22.5 GW on the Yangzi) that is 185 m tall and 2.3 km long and contains nearly 28 million m^3 of concrete reinforced with nearly half a million tonnes of steel bars (Smil, 2013). China's domestic demand has not been able to absorb all of the country's huge steel production, and this has led to higher exports, falling prices, and rising inventories. Already in 2005 the country's surplus steel capacity was estimated at 80 Mt (Okuno, 2006), and the Chinese exports (in net terms) rose from 22 Mt/year in 2010 to 43.4 Mt in 2013 and surpassed 60 Mt in 2014.

Figure 4.5 Reinforcing bars on a construction site in Nanjing (Jiangsu province) in China. *Corbis.*

And it has not been just ordinary rebar or wire; Chinese steel is now embedded in many of the world's most prominent structures. After the San Francisco–Oakland Bay Bridge (originally opened to traffic in November 1936) was damaged by a 1989 earthquake, the 2002–2013 reconstruction of the eastern section in the form of a self-anchored suspension bridge (the structure's visual dominant) became a Chinese affair. The bridge's 28 individual steel deck sections with seismically resistant hinged expansion joints were made by Baosteel in Shanghai, and so was the main cable (the world's longest looped suspension bridge cable with 17,399 compressed wires) and 158 m tall steel tower (AIG, 2015).

Since 2005 the Chinese government has been promising to reduce the country's excess capacity, and yet it has persisted. Excess capacity is now a worldwide phenomenon created by the combination of stagnant or declining steel demand in most of the affluent countries and by the precipitous post-2000 expansion of Chinese steelmaking. This has been true not only for the common low-quality categories of steel, but also for stainless steel. Long-standing European overcapacity (EU produces more than twice as much steel as North America) and Chinese over-expansion (in 2007 China produced less than 7 Mt of stainless steel a year; by 2012 the output was above 16 Mt) created the prospect for subdued global growth (Millbank, 2013).

But there has been a major shift in Chinese expectations as the extraordinarily fast expansion of the post-2000 steel production made an inevitable slowdown come much sooner than anticipated by forecasts made just 10–15 years ago. During 2014 Chinese steel demand began to shrink for the first time since the year 2000, and steel prices have fallen to such an extent that by October 2014 one of the most quoted news items presented by Chinese Central Television was the claim that "steel is now almost as cheap as Chinese cabbage," the cheapest of all of the country's common green vegetables (CCTV, 2014). But the falling prices failed to stimulate domestic demand, profit margins are meager (on the order of 0.5%), and debt burden has increased (by June 2014 it totaled nearly half a trillion dollars), as have inventories of iron ore and finished steel products.

At the same time, China continued to import high-quality metal for various exacting applications and tried to reduce future steel demand (and also excessive levels of borrowing) by enacting bans on new construction projects. This is an apposite place to note that Sweden is the very opposite of China's mass production of basic steel products: about 65% of Swedish steel output are special alloys (compared to about 15% in Japan and less than 10% in the United States), and such relatively small companies as Sandvik, Böhler-Uddeholm, Ovako, and Höganäs are the world's leading producers of, respectively, seamless tubes, tool steel, ball-bearing steel, and iron and steel powder. The country exports about 95% of its high-quality production (Jernkontoret, 2014).

CHAPTER 5

Modern Ironmaking and Steelmaking

Furnaces, Processes, and Casting

In this chapter, I will first review the state-of-the-art performance of modern ironmaking, both by the world's largest blast furnaces and by the direct reduction of iron (DRI). The DRI processes represent a fundamental ironmaking innovation of the post-WW II era, but, although they have enjoyed some commercial success and are now operating in two dozen countries, these new metallurgical techniques have not fulfilled their early promise: by 2015 only about 5% of the global iron output originated in DRI plants. In contrast, the state-of-the-art processes of modern steelmaking include two processes that were introduced only during the 1950s but whose worldwide adoption has been (a few notable exceptions aside) extraordinarily rapid.

Without exaggeration, swift diffusion of basic oxygen furnaces (BOFs) and of continuous casting have revolutionized the industry through higher efficiencies, reduced waste, and rising productivity. The third major pillar of modern steelmaking, smelting of recycled scrap metal in electric arc furnaces (EAFs), is a process that has been practiced for more than a century, but only in recent decades has it become more than an also-run compared to open hearth furnaces (OHFs) or BOFs: it is now the dominant (more than 90%) steelmaking process in the Middle East (using scrap and DRI), Africa (nearly 70%), and in the United States (about 60%), and it produces nearly 30% of the world's crude steel (WSA, 2015). In this chapter, I will review technical advances and the best recent performances of BOF and EAF smelting and continuous casting, and in the next chapter, I will focus on material flows of iron- and steelmaking processes and on their energy requirements.

NEW BLAST FURNACES

Development of modern blast furnaces during the past 100 years can be followed in books by Boylan (1975), Boylston (1936), Geerdes, Toxopeus, and

Still the Iron Age.
DOI: http://dx.doi.org/10.1016/B978-0-12-804233-5.00005-1

van der Vliet (2009), Hogan (1971), King (1948), Lovis (2005), and Peacey and Davenport (1979). An illustrated look at their construction, charging, and ancillary operations (hot blast stoves, casting, slag processing) is offered by one of their leading builders (Siemens VAI, 2008). A unique tribute to their former glory was assembled by Becher and Becher (1990): the couple had traveled widely in the regions of American and European ironmaking—Pennsylvania, Ohio, Indiana, Alabama, Ruhrgebiet, Saarland, Lorraine—to photograph more than 200 blast furnaces, both working and abandoned, in order to document the now classic phase of Western industrialization.

After more than a century of growth, modern blast furnaces reached their capacity plateau between 1973 and 1980. As already noted, the world's largest blast furnace (No. 2 furnace at Nippon Steel Oita Works) was blown-in in 1976. Its start-up was preceded (on February 13, 1973) by the completion of Europe's largest blast furnace, and it was followed (in October 1980) by the blowing-in of America's largest furnace. Europe's record holder was Schwelgern 1 at August-Thyssen Hütte in Duisburg-Marxloh, 33 m tall, with a hearth diameter of 14 m, volume of 4200 m^3, and daily output of 10,000 t (ThyssenKrupp, 2003). America's largest furnace, Inland's No. 7 (East Chicago, Indiana), built by Koppers, had a hearth diameter of 13.5 m, inner volume of 4758 m^3, working volume of 3470 m^3, and daily output of about 9000 t of pig iron (McManus, 1981). This compared to an average working volume of 1600 m^3 at US Steel and about 1800 m^3 at Bethlehem Steel (McManus, 1988a). By that time only the Japanese ironmakers had the experience with operating such large furnaces, and the planning and early operation of No. 7 benefited from the advice of Nippon Steel.

When these furnaces were planned and built, nobody suspected what reverses lay ahead for Western and Japanese ironmaking, and that these structures would remain record holders for decades to come. OPEC's first oil price rise in 1973–1974 was followed by the second round in 1979–1981, resulting in record high prices of oil and increased prices of coal, natural gas, and electricity and causing a worldwide economic slowdown. As a result, falling pig iron production—in Japan from 90 to 73 Mt between 1973 and 1983 and in the United States from nearly 92 to just 44.2 Mt during the same 10 years—had not only put an end to the building of ever larger furnaces but had also led to the closure of many older units (Becher & Becher, 1990). Eventually some aging furnaces—including the three record holders—were rebuilt or relined and relaunched as more efficient smelters with more durable linings (graphitic materials) and more efficient and hotter blast stoves.

Inland's No. 7 was shut down for relining in August 1987 and by 1988 the US industry was relining or rebuilding nearly 40% of its remaining active furnaces (McManus, 1988a). And hence when in 1989 a long paper in America's leading ironmaking periodical asked "Is the blast furnace in its twilight?" (Hess, 1989), the answer could be not yet, as continuing technical improvements and major relining and rebuilding efforts guaranteed to keep the world's largest ironmakers operating well into the next century. But a decade later the doubts were back: in 1997 no full-size blast furnace was ordered, and in 1998 only China returned to building them—but, once again, the expert consensus was that the demand for pig iron, the global scale of ironmaking, and the progress on alternative iron plants would support blast furnaces for decades to come. By 2015, that verdict was only strengthened, not only because of the intervening expansion of South Korea's and China's ironmaking but because cumulative advances and major reconstructions have been giving the world's largest blast furnace an even longer lease on life.

After its second campaign of 6 years, the Inland Steel's No. 7 was relined by 1200 workers of Edward Gray Corporation in just 29 days in 1993, and its third relining, completed in March 2004, set it for a 20-year campaign (Gary Works is now a part of ArcelorMittal). The most recent relining of Schwelgern 1 with 5500 t of advanced refractory materials took place in early 2008 after the fourth campaign of about 12 years (ThyssenKrupp, 2008), and during the first 40 years of its operation it smelted 115 Mt of pig iron (WAZ, 2013). In 1993, a larger Schwelgern 2 (14.9 m hearth diameter, 4800 m^3) was blown-in. It had produced 78 Mt of pig iron during its first campaign of 21 years, and after relining it began its second campaign in October 2014 (ThyssenKrupp, 2014).

After its first campaign, Oita No. 2 was enlarged from 5070 to 5245 m^3 and then averaged nearly 11,200 t of hot metal a day, and on August 8, 1997, it set a new record of 13,368 t. After its second campaign, Nippon Steel rebuilt the furnace in 2004 in just 79 days. The reconstruction used a hydraulic system and piping by Kawasaki Precision Machinery, and it enlarged every part of the structure, setting a new world record: the inner volume from 5245 to 5775 m^3, working volume from 4312 to 4753 m^3, hearth diameter from 14.9 to 15.6 m, belly diameter from 16.6 to 17.2 m, number of tuyères from 40 to 42, and number of tapholes from 4 to 5 (Haga, 2004).

The furnace was blown-in on May 15, 2004. Just a month later the furnace was producing 12,500 t a day, and 5 months later it reached its designed capacity of 13,500 t a day.

In 2009 Oita No. 1, first blown-in in 1972, was enlarged to the same volume as the second furnace (5775 m^3), but in that year Japan had narrowly lost its long-held record with the 2009 completion of the Zhangjiagang II No. 4 blast furnace of Shagang Group, China's largest privately owned company in Jiangsu province: with a volume of 5800 m^3 and hearth diameter of 15.7 m, it was the first furnace with nominal capacity of 5 Mt per year and with actual output of 4.8 Mt (WISDRI, 2012). But its record rating lasted only until 2013: in June of that year South Korea's POSCO renovated (in 108 days) its Gwangyang No. 1 furnace in Gwangyang Steelworks in South Jeolla province. The furnace, built originally in 1987 (3800 m^3), was later slightly enlarged to 3950 m^3: the second enlargement brought it to 6000 m^3, with a hearth diameter of 16.1 m and annual capacity of 5.46 Mt (POSCO, 2013).

By 2013, there were 21 blast furnaces with capacities of 5000 m^3 or more, with 12 of them at or above 5500 m^3: Gwangyang No. 1, Zhangjiagang II No. 4, Oita No. 1 and No. 2, Pohang No. 4 (POSCO, 5600 m^3), Cherepoverts No. 5 (Russia's Severstal, 5580 m^3), Caofeidian No. 1 and No. 2 (China's Shougang Jingtang Iron & Steel, both at 5576 m^3), Kimitsu No. 4 (NSSMC, 5555 m^3), Schwelgern No. 2 (ThyssenKrupp, 5513 m^3), Gwangyang No. 4 (POSCO, 5500 m^3; in 2010 it achieved the world record output of 15,613 t/day), and Fukuyama No. 5 (JFE Steel, 5500 m^3). Asian lead was obvious: 10 of the 12 largest furnaces were in Asia (4 in Japan, and 3 each in South Korea and China), just 1 was in Europe, and none were in North America.

In 2013, 287 large (>2000 m^3) furnaces operated worldwide, compared to 234 furnaces with volumes between 1000 and 2000 m^3 and 295 furnaces smaller than 1000 m^3 (VDEh, 2013). Notable concentrations of very large blast furnaces include Baoshan (Shanghai), Anshan and Benshi (both in Liaoning), Caofeidian and Qian'an (both in Hebei) in China, Yawata in Japan, Pohang in South Korea (Fig. 5.1), Gary (Indiana) in the United States, Cherepovets in Russia and Duisburg, and Nordrhein-Westfalen in Germany. Europe's largest iron-smelting center is in Schwelgern, just north of Duisburg on the eastern shore of the Rhein, where ThyssenKrupp Steel Europe and Hüttenwerke Krupp Mannesmann GmbH operate six blast furnaces (with hearth diameters of 10.0–11.9 m) producing annually up to 15.6 million tonnes of hot metal (ThyssenKrupp, 2015; Fig. 5.1).

Figure 5.1 POSCO's steel mill in Pohang, North Gyeongsang province, South Korea (top), and an aerial view of the ThyssenKrupp iron and steel mill in Schwelgern, on the eastern shore of the Rhine, north of Duisburg, in Nordrhein-Westfalen, Germany (bottom). *Corbis.*

Post-1990 reconstructions tended to have slightly higher capacities even for medium-sized furnaces: for example, nine furnaces relined at Nippon Steel plants between 1990 and 2004 had an average 14% increase of inner volume to 4490 m^3 (Kawaoka et al., 2006). And better linings (graphitic refractories with cooling plates) have resulted in unprecedented extensions of typical smelting campaigns, leading to doubling, or even tripling, of the time spans between reconstructions. During the early 1980s, campaigns lasted usually no longer than 3–5 years, by the century's end campaigns of 8–10 years were common, and the record holder at that time was OneSteel's Whyalla furnace in Australia, which was relined in 1980 and had a planned overhaul scheduled for 2004 after a generation-long campaign (Bagsarian, 2001).

The complete record for the Nippon Steel furnaces shows that the campaigns of the units blown out during the 1970s (first campaigns of Nagoya 3, Kimitsu 2, and Oita 1) were just 5–7 years; those blown out during the 1980s and 1990s lasted roughly twice as long, 10–12 years, and the blast furnaces blown out just before and after the year 2000 (including the second campaign of Oita 2) lasted about 15 years without losing their high productivity (Kawaoka et al., 2006). Even China's best furnaces could now last that long: for example, due to reducing the progress of two main problems, breakouts of hearth walls and the erosion of shaft lining, the fifth furnace of the Wuhan Iron and Steel Company, blown-in in 1991, worked smoothly for 15.6 years and produced 35.51 Mt (i.e., 11,096.6 t/m^3) without any repairs (Zhang & Yu, 2009). And, as just noted, the first campaign of Schwelgern 2 lasted 21 years.

Shinotake et al. (2004) analyzed the relationship between campaign length and furnace productivity and found that furnaces whose cumulative productivity was less than 10,000 t/m^3 had campaigns of 12–18 years, while those whose production reached 14,000 t/m^3 lasted 18–25 years. In the past, major problems tended to arise in the shaft, while the longevity of more recent campaigns is usually controlled by the state of the hearth. Measures to prolong the service include the installation of highly conducive cast iron or copper stave for the hearth sidewall below tapholes, lower temperature of cooling water, two-step cooling for hearths, and durable carbon blocks and inner ceramic linings for the bottom refractory. Better linings and better stoves—the latest ones having readily identifiable mushroom domes (Siemens VAI, 2008)—have made it possible to raise hot blast temperatures to 1250 °C (maximum dome temperature of 1400 °C) and to increase average daily output by 15–25%. Newly relined large furnaces usually produce between 10,000 and 12,000 t of hot metal a day, and they do so with reduced energy requirements, with major coal injection rates, and, in some furnaces, due to injection of plastic waste.

But the largest furnaces with the highest absolute production do not have the highest specific metal output: the rates for furnaces with inner volumes of 5000 m^3 and above and with daily production of at least 10,000 t (and up to more than 12,000 t) range between 2.1 and 2.7 tonnes per day per m^3. In contrast, furnaces with volumes between 1800 and 2700 m^3 and daily maxima of 6000–9000 t have higher specific daily rates in excess of 3 t/m^3. Dutch Ijmuiden No. 6 (belonging to Tata Steel Europe, volume of 2678 m^3 and daily output of 7800 t) rates as high as 3.35 tonnes per day per m^3 (VDEh, 2013).

Injection of pulverized coal is a method that was first used by Armco in Ashland, KY, in 1963, but (with the exception of Shougang mill in China) it found little acceptance elsewhere as long as inexpensive oil and natural gas could be used as supplementary fuels to replace part of the coke charge, a situation that changed after two OPEC-led oil price increases (McManus, 1989). Nippon Steel licensed the Armco process in 1981 as the record high oil prices led to a rapid Japanese adoption of coal injection: by 1986 half of the Japanese blast furnaces used it. The first British Steel installation came in 1982, and by the late 1980s there were nearly 50 furnaces using the process, replacing up to 40% of coke with coal (at 1:1 rate), whose cost was only 35–45% as much per unit of weight. During the 1980s, most furnaces received less than 100 kg/t, but during the 1990s many furnaces began to work with as much as 175 kg of coal per tonne of hot metal, and in 1998 US Steel Gary Works began testing simultaneous injection of natural gas and pulverized coal, replacing unprecedented shares of coke.

The cooling effect of injecting pulverized coal makes it possible to use higher hot blast temperatures and higher concentrations of injected oxygen, and this results in reduced total fuel consumption and higher furnace productivity (Danieli Corus, 2014). Nomura and Callcott (2011) made a theoretical investigation of the maximum pulverized coal injection rates and concluded that they are between 190 and 210 kg/t of hot metal. Although some large furnaces operated for short periods of time with higher injection rates (up to about 260 kg/t), totals close to 200 kg/t are in agreement with the highest reported rates used for long periods of time in stable blast furnaces. They found that the replacement of coal by coke increases the generation of CO and H_2 and lowers the bosh gas temperature. Coal injection has been a superior option to investing in new coking batteries, but there are clear limits both because of the structural support and permeability provided by coke as well as due to the fact that high rates of coal injection require the enrichment of the hot blast with oxygen, offsetting the coke savings.

In 1996, NKK began the injection of coarse grains of used plastics as an alternative to coal at its No. 2 Keihin Works furnace south of the capital. The company was charging annually as much as 120,000 t of shredded plastic per blast furnace, saving 1.1 tonnes of coke per tonne of plastics, reducing annual energy use by about 1.5%, and reducing CO_2 emissions by about 3.5 kg C/t of hot metal (Ogaki et al., 2001). JFE Steel (established in 2003 by the merger of NKK and Kawasaki Steel) continues the practice at Keihin Works (Fig. 5.2). Nippon Steel Corporation had also developed a new process for waste plastic recycling in coke ovens (Kato et al., 2006).

Figure 5.2 JFE's No. 2 blast furnace at Keihin Works, south of Tōkyō. *Reproduced by permission from JFE Steel.*

And the late 1990s also saw the first trials of direct hot oxygen injection (as opposed to further oxygen enrichment of the hot blast) into blast furnaces: Praxair developed a thermal nozzle injecting high-temperature oxygen, and this technique should improve coal burnout by increasing oxygen levels in the vicinity of the injected powdered coal plume (Halder, 2011). Another source of increased efficiency, to be described in some detail in the next chapter, has been the charging of beneficiated (sintered or pelletized) ores.

DIRECT REDUCED IRON

Before moving to a brief review of state-of-the-art steelmaking practices, I must devote a few pages to describing the only major practical alternative to BF ironmaking, direct reduced iron (DRI). This group of techniques obviates the use of metallurgical coke as it reduces iron ores in their solid state at temperatures well below the metal's melting point. In principle, DRI could be thought of as a modern, efficient replication of the preindustrial, artisanal production of spongy iron masses in bloomeries, the process described in this book's first chapter. Between 1869 and 1877, William Siemens experimented with a variant of DRI by attempting to reduce a mixture of crushed high-quality iron ore and coal in rotating cylindrical furnaces, and during the 1920s two Swedish processes were employed in a small-scale local production of iron powders. One of them (Höganäs process) is still used for that purpose by an eponymous Swedish company and by other enterprises (Höganäs, 2015).

Commercialization of DRI began only during the late 1960s, and by the mid-1970s there was a choice of nearly 100 designs combining different reactors (furnaces, kilns, retorts, fluidized-bed reactors) with a number of reducing agents, including coal, graphite, char, liquid and solid hydrocarbons, and gases (Anameric & Kawatra 2015; Hasanbeigi, Price, & Arens, 2013). DRI processes can be classified according to the kind and source of reducing gas or type of reactor used, and the DRI process that has been so far the most commercially successful relies on natural gas and hence it has been most commonly installed in locations and in countries where this fuel has been inexpensive and readily available. Natural gas is reformed by a catalytic steam process ($CH_4 + H_2O \rightarrow CO + 3H_2$) and it reduces iron pellets or fines as it ascends, mixed with crushed limestone, in a shaft (Anameric & Kawatra, 2015). The Mexican *Hojalata y Lamina* was the pioneering design (with later versions marketed as HYL-III), and the most successful US contribution, MIDREX process, uses shaft furnaces, while Fior/Finmet, Iron Carbide, and Circored use fluidized-bed reactors.

MIDREX has earned its leadership in direct reduction because its plants are the industry's most productive and most reliable (often operating for more than 8000 hours a year) and can use a range of reductants and raw materials (MIDREX, 2015). Plants can produce reducing gas from the energy source that is either locally, or most readily, available or that is most competitively priced, be it natural gas (reformed to yield CO), syngas from coal, petroleum coke ore heavy refinery residue (processed in a gasifier), or coke oven gas. Furnaces can work either with lump ore or iron oxide pellets (or their mixture), and the process itself is simple (involving the countercurrent descent of iron-bearing material loaded at the top of a cylindrical vessel lined with refractories and ascent of reducing gases).

MIDREX DRI is in the form of a solid sponge, vulnerable to reoxidation and unsuitable for transportation, and since 1984 this sponge iron has been first converted into hot-briquetted iron (HBI), produced by discharging hot DRI into roller presses that mold it into dense, pillow-shaped briquettes highly suitable as EAF charge. The other products are cold DRI (cooled before discharging) and hot DRI in the form of dense (bulk density of 2.5–3 t/m^3) briquettes containing 90–94% Fe and each weighing up to 3 kg. When compared to blast furnace smelting, the process has three other key advantages besides doing away with coke. First, its specific energy requirement (GJ/t) is approximately only half of those for BF operation (for details see the penultimate section of the next chapter). Second, as a result, its specific carbon emission is also much lower (see the last section of the next chapter). Third, there is a much greater flexibility in designing plant capacities: while the economies of scale for the standard BF–BOF choice demand annual output of at least 2 Mt, DRI plants associated with a mini-mill can produce as little as 0.5 Mt/year.

Corex, the most successful smelting-reduction process, was originally developed by *Voest-Alpine Industrieanlagenbau* (VAI), and it is now marketed by Siemens under the name SIMETAL Corex (Siemens VAI, 2011). It requires two separate process reactors, the first being a reduction shaft charged with lump ore or pellets and additives to produce DRI in a counterflow reaction, and the second being a melter gasifier where the reduction is completed and hot metal and slag are tapped, much as in conventional BF. The first Corex plant began to operate in Pretoria in 1989, and subsequent installations included Saldanha in South Africa, Pohang in

South Korea, at China's Baosteel (two units), and five units in India (JSW Steel and Essar Steel).

Rotary hearth furnaces (RHFs) have been used for decades in heat treating metals and high-temperature recovery of nonferrous metals, and hence their use in ironmaking has been a matter of specific application and appropriate process control. RHFs—developed and marketed under proprietary labels of Fastmet, Fastmelt, Redsmelt, Sidcomet, Primus, and ITmlk3—now constitute the largest class of new direct reduction processes but account for a minority of global DRI output (Anameric & Kawatra, 2015; Guglielmini & Degel, 2007; McCelland, 2002; Sohn & Fruehan, 2006). Their flat refractory hearths rotate inside high-temperature circular tunnel kilns lined with refractories, with iron ore and reductant in a single- or multi-layer bed, and with temperature controlled by burners (using natural gas, fuel oil, or pulverized coal) along the wall and on the roof. Mixed ore-and-coal pellets are subjected to temperatures up to 1300°C, mostly for just 6–12min.

The product is not liquid and, inevitably, it contains relatively large amounts of ash and nonmetallic residues from the processed coal–ore mixture, and it must be melted in EAF. The process requires constant feeding and has limited productivity. The first commercial RHF, designed to recycle wastes containing Ni and Cr at INMETCO in Elwood, PA, went into operation in 1997 (INMETCO, 2015). MIDREX began developing an RHF process concurrently with its countercurrent reactor, using reformed natural gas, abandoned the quest in favor of the latter technique, and returned to it by 1992 with the development of the Fastmet process. The first Japanese plant has been operating since the year 2000 at Hirohata Works, and there are now six plants with combined annual capacity of nearly 1Mt (Kobelco, 2015a). The process is particularly suitable for converting such iron- and steelmaking wastes as blast furnace dusts and sludges and EAF dust and mill scale.

Obvious drawbacks are the need for constant feeding and a limited productivity. Fastmelt is Fastmet with an added electric iron melting furnace to produce hot DRI. Currently the most advanced RHF design (which can be seen as a variation of Fastmet) is the ITmk3 (Ironmaking Technology Mark 3, BF being the first, and gas-fueled DRI the second generation) process developed by Kobelco Steel (Harada & Tanaka, 2011; Kobelco, 2015b).

Iron ore concentrate (magnetite or hematite or their mixtures) and noncoking coal mixed in pellets are processed for 10min by a single-stage heat treatment in an RHF, yielding high-quality (96–97% Fe), slag-free iron nuggets. These nuggets are easier to handle than DRI or hot-briquetted iron and are processed in EAF or (after remelting) in BOF. Because nearly all combustible gases generated by the reactions are burnt within the furnace, the process can reduce specific CO_2 emissions by 8–17% compared to BF. The first plant, at Hoyt Lakes, MN, began operating in 2010 and it had an annual capacity of 500,000 t.

Not surprisingly, during the early years of its commercialization DRI received enthusiastic reception and there were high expectations for its continuing strong expansion, if not an eventual dominance of primary iron production. Global DRI production rose nearly 10-fold during the 1970s (to 7.1 Mt), and Miller (1976) forecast the output of 120 Mt by 1985—but the global capacity in that year was just over 20 Mt, and actual sales were only 11 Mt (MIDREX, 2014a). Expectations shifted, as slow but steady growth came to represent DRI success. Global output rose to 31 Mt in 1995, 43 Mt in 2000, and 75.22 Mt in 2013, and DRI's share of the global primary iron production (excluding the output of RHFs using recovered mill wastes) rose from 1% in 1980 to 4.7% in 2013.

In 2013, there were more than 120 DRI plants worldwide and MIDREX (now owned by Kobe Steel) remains the premiere DRI licensor as its plants have been supplying 60% of all DRI for more than two decades and produced 63% of all DRI in 2013 (MIDREX, 2014b). In 2013, the largest of the 56 plants were in Mobarakeh in Iran (capacity of 3.2 Mt in five modules) and in Corpus Christi, TX (2 Mt). The other leading shaft reduction process (HYL/Energiron) supplied about 15% of all DRI in 2013, while coal-based RHFs delivered about 21%. Maximum annual capacity of MIDREX installations grew from less than 0.5 Mt (Series 400) to more than 3 Mt (MEGAMOD series), and the oldest MIDREX reactors have been in service for more than 40 years (MIDREX, 2014b). The Middle East is the leading (natural gas-based) producer with nearly 40% of the global total; India, with more than 40 (mostly coal-based) plants, is the leading nation (nearly 25% of the total), and both the United States and the EU are only negligible producers.

BASIC OXYGEN FURNACES

Because open hearth furnaces dominated global steel production until the 1960s, their capacities had to increase to accommodate the rising demand. The largest furnace owned by US Steel in 1910 had a hearth area of about

40 m^2 and capacity of 42 t, while the record-sized furnace introduced during WW II had a hearth of 85 m^2 and produced 200 t of steel in a single heat (King, 1948).

OHFs had benefited from the post-WW II expansion of steel production when they reached their greatest scale in both the United States and in the USSR, with new units capable of producing more than 600 t of steel in a single heat, and the world's largest furnace in the USSR had a capacity of 900 t. But by the time it went into operation the dominance of OHFs was over: in the early 1950s, it appeared that they would remain in demand for decades to come, but by the end of the twentieth century they produced just 4% of all steel, with Russia and Ukraine being the only two major producers with high shares of open hearths, respectively 27% and 50% (WSA, 2001).

This was the result of a veritable technical revolution, but because it unfolded during the same time as the advances of the electronic world (that have always received disproportionate media attention), its impressive achievements have received hardly any public appreciation. As a result, Ohashi (1992, p. 540) had to conclude that

> *in the popular conception, steelmaking is the quintessential example of an outmoded manufacturing technology, a 19th-century dinosaur that is about to lumber obliviously over the threshold of the 21st century. But in fact, steelmaking has been transformed in the past 20 years by a flood of innovations. With little fanfare, it has become as impressive as that acme of modern manufacturing practice, integrated-circuit processing.*

This flood of innovations, which brought the end of the open hearth furnace era, began during the early 1950s when the European steelmakers pioneered the adoption of BOFs, fundamentally nothing but improved variants of the old Bessemer converter: there is no external source of heat and the process relies on the exothermic reaction between iron and oxygen. In fact, Bessemer obtained a British patent (UK 2,207, on October 5, 1858) that envisaged the blowing of oxygen to decarburize iron. That idea could not be transformed into commercial reality as long as oxygen was not available in large volumes and at an affordable cost. The road to that was opened by Mathias Fränkl's tubular regenerator (heat exchanger), patented in 1928 and incorporated into the Linde-Fränkl process, whose first commercial installation came in 1934. New regenerators were very efficient, and they were a major reason why the price of oxygen produced by the new Linde process offered to industrial consumers in 1950 was reduced by an order of magnitude.

As is the case with Bessemer or open hearth furnaces, the adjective basic does not refer to any essential quality or the simplicity of design but to the pH level of the slag: magnesium oxide (MgO) in the furnace's lining is used to remove and retain trace quantities of P and S from the molten metal. Added slag also prolongs the lining's life, and after every competed heat the vessel is rocked to distribute the slag up its sides. The development and commercialization of BOFs were remarkable for the absence of any of the world's major steelmakers: they were driven by the dedication of a Swiss metallurgist and the foresight of managers in two small Austrian companies.

After Swiss metallurgist Robert Durrer (1890–1978) graduated from the Aaachen University in 1915, he remained in Germany, and in 1928 he became a professor of *Eisenhüttenkunde* at Berlin's *Technische Hochschule*, where he began to experiment with oxygen in steel production. When he returned to Switzerland in 1943, he joined the board of the country's largest steel company, von Roll AG, and, helped by German metallurgist Heinrich Hellbrügge, he continued his research at a steel plant in Gerlafingen (Starratt, 1960). As is often the case with major inventions, Durrer had competitors: C.V. Schwarz obtained a German patent (735,196) for a top-blown oxygen converter in 1943, and John Miles received a Belgian patent (468,316) in 1946 (Adams & Dirlam, 1966).

But neither of these ideas was translated into a working prototype before Durrer made a breakthrough in April 1948 when a small converter bought during the previous year in the United States was used to produce steel by blowing with pure oxygen. Moreover, Durrer and Heinrich Hellbrügge, the metallurgist in charge of the experiment, established that cold scrap could amount to as much as more than half of the hot metal weight. Soon afterwards, Hellbrügge notified Herman Trenkler at VÖEST (*Vereinigte Österreichische Eisen- und Stahlwerke AG*, the country's largest steelmaker) that the oxygen process was ready to be tried in a commercial setting, and von Roll, VÖEST, and Alpine Montan AG quickly concluded an agreement for such testing at VÖEST's Linz plant and at Alpine's Donawitz plant.

Initial operation problems were overcome in a matter of months; construction of two 30-t oxygen furnaces began in December 1949, and VÖEST produced its first steel on November 27, 1952, and Alpine in May 1953 (Starratt, 1960; Stubbles, 2015). That is why the new method became commonly known as the Linz-Donawitz process (*das LD-Verfahren*) and why many people considered it, wrongly, an Austrian invention. Given its intellectual and practical origins, the method should be known as the

Bessemer–Durrer process. VÖEST eventually became the sole proprietor of the L–D process, but the world's major steelmakers were in no hurry to license new oxygen furnaces from a small Austrian company whose aggregate capacity was less than a third of the output of a single plant of US Steel (Adams & Dirlam, 1966). Although the advantages of the process were widely recognized by US steelmakers, corporate inertia among large companies and their high sunk costs in open hearth furnaces meant that all major US producers stayed far behind the oxygen wave (ASME, 1985; Emerick, 1954; Hogan, 1971).

McLouth Steel in Trenton, MI, a small company with less than 1% of the country's ingot capacity, became the technique's US pioneer in late 1954, capable of producing about 490,000 t a year from three 60-t furnaces with 45-min heats (ASME, 1985). The next installations came only in 1958 (two 100-t furnaces at McLouth and three 100-t furnaces at Kaiser Steel in Fontana, CA), and the two leading US producers, US Steel and Bethlehem Steel, began to operate their first BOFs only in 1963 and 1964 (Hogan, 1971). Adams and Dirlam (1966) used this much belated adoption as the principal example to argue against the universal validity of the Schumpetrian hypothesis, according to which large firms with great market power have greater incentives and more abundant resources to be the innovation leaders (Schumpeter, 1942).

The adoption pace finally quickened during the 1960s, and by 1970 oxygen furnaces produced 40% of American steel. But Japanese steelmakers, rebuilding the plants destroyed during WW II, embraced the technique so rapidly that by 1970 80% of the country's steel production originated in BOFs. Subsequent steady progress brought the global share to nearly 60% by the century's end, when the US and Japanese shares were, respectively, a bit above 50% and more than 70%, and Germany produced 70% of its steel in BOF and Austria 90%. By 2013, the global share reached 70% and many national shares changed only a bit, but the US share declined to about 40%.

Modern BOFs are massive, open-mouthed, pear-shaped vessels that require 50–60 m^3 of oxygen to produce 1 t of crude steel; the gas is blown from both the top and the bottom, as a supersonic jet from a water-cooled vertical lances onto molten pig iron and through the tuyères at the furnace bottom, which are also used to blow in argon (an inert gas) to stir the charge and reduce phosphorus levels in the metal (Miller et al., 1998; Fig. 5.3). The resulting oxidation is much more rapid than in open hearth furnaces: a large BOF decarburizes 300 t of iron from 4.3 to 0.04% C in

Figure 5.3 Molten pig iron is poured into a BOF to make steel at JFE Steel's Kimitsu Works in Chiba, east of Tōkyō. *Reproduced by permission from JFE Steel.*

just 35 min, while the most advanced open hearth would need at least 9 h. The entire process is slightly exothermic (hot gas leaving the furnace, typically 90% CO and 10% CO_2, is often used to melt the charged scrap), yielding about 200 MJ/t of crude steel. The largest furnaces are 10 m high and up to 8 m in diameter, and their capacities range mostly between 150 and 300 t/heat.

Productivity advances have been related primarily to the use of oxygen: the early norm was less than 9 m^3 of oxygen per tonne of steel; by the late 1990s the typical rate rose to 30 m^3/t. In the year 2000, the world's largest BOF, with capacity of 375 t and tap weights of 325 t in 94 min, was at Northwestern Steel & Wire Company in Sterling, IL.

In the latest furnaces, oxygen use is up to 50 m^3/t, typical inputs per tonne of steel include 1033 t of hot metal, about 40 kg of scrap and ferroalloy, and about 30 kg of flux (lime and dolomite), and by-products include about 100 m^3/t of hot gas (about 70% CO, 10–15% CO_2) and 50 kg of slag. Charging a furnace with pig iron (delivered by torpedo cars, poured into a ladle, and delivered by an overhead crane) requires tilting of the furnace on its pinions. After it is charged the furnace is returned to the upright position, liquid-cooled lances are inserted, blowing with a high-purity oxygen begins, and slag-forming material is added. About two-thirds of the required oxygen converts C to CO, and 9% goes for oxidizing Si to SiO_2 and another 9% for turning Fe to FeO in slag. After a heat is finished the molten steel is transferred to a ladle and either delivered to a tundish to begin continuous casting (see the following section) or cast into ingots.

The economic impact of BOFs was truly revolutionary. Capital expenditures were cut by an order of magnitude as a single BOF can replace 10–12 open hearth furnaces, and labor productivity rose by 3 orders of magnitude, from 3 man-hours/t of hot metal produced by open hearths in 1920 to 10 s/t by BOFs in 1999 (Berry, Ritt, & Greissel, 1999; Shepard, 2004). Reduced flexibility of charges has been the only drawback: scrap could make up as much as 80% of the open hearth charge but it is limited to no more than 30% in oxygen furnaces. But the unchallenged dominance of these furnaces was much shorter than the reliance on open hearths: by the end of the twentieth century their share of steel output was declining in every major steel-producing nation because the share of the metal produced in EAFs was rapidly increasing (see Fig. 9.5).

ELECTRIC ARC FURNACES

Origin of electric arc furnaces goes back to the experiments conducted in 1878 and 1879 by William Siemens, one of the eminent inventors and innovators of the nineteenth century, whose other notable metallurgical innovation, the regenerative furnace, was described in the previous chapter. Siemens built two furnaces, one with electrodes at the top and at the bottom (charged material covered the lower electrode), and the other with horizontally opposing electrodes melting the charge beneath them by radiation. But during the late 1870s, there was no way to generate electricity on large scale and at an affordable cost (the first commercial coal-fired power plants began to operate in 1882), and the first commercial use of relatively small EAFs came in 1888 for the newly invented process of smelting aluminum invented by Charles Hall (1863–1914) and Paul Héroult (Smil, 2005; Toulouevski & Zinurov, 2010).

Before WW I, arc furnaces were used (in small numbers) in places with cheap hydroelectricity as smelters (Noble Electric Steel in California, Stassano furnace design in Torino), and they were charged with pulverized iron ore and charcoal, but most of them were installed as replacements of Bessemer converters or open hearths. By 1913, Héroult's furnaces produced about 750,000 t of steel, with Germany accounting for 50% and the United States for nearly 20% (*Scientific American*, 1913). Furnaces were used to make small batches of specialty steel, and their capacities gradually increased from less than 5 t before 1910 to about 25 t by the late 1920s. American electric arc steel production grew by nearly 90% during the 1920s, then dropped and stagnated during the 1930s, surpassed 1 Mt in 1940, and tripled by the end of WW II as the largest furnace capacities reached 100 t/heat.

EAFs gained a large share of steel production only during the great post-WW II industrial expansion when electricity prices declined and when the previous decades of iron and steel production resulted in accumulation of relatively large amounts of scrap metal. Another key factor in their rapid adoption was the lower capital cost of these furnaces, no more than 15–20% of the total cost of operation combining blast furnace and BOF (Jones, Bowman, & Lefrank, 1998). Key technical advances since the middle of the twentieth century included larger capacities, shorter tap-to-tap times (hence higher productivities), and reduced energy and electrode-graphite consumption.

By the end of the twentieth century, the best melt shop, at Badische Stahlwerke in Kehl, set a new world record with its best furnace producing 46 heats of 87 t, with tap times down to 36 min (power-on time less than 28 min), average electricity consumption of 302 kWh/t, and oxygen use of 37.5 m^3/t (Greissel, 2000). Typical performances have improved as follows: tap-to-tap times declined from 3 h in 1970 to just 30–40 min three decades later, and specific electricity requirements fell from more than 700 kWh/t in 1950 to 475 kWh/t by 1980 and then to less than 350 kWh/t by 2010, while graphite consumption was reduced from about 7 kg/t in the early 1960s to as little as 1.1 kg/t by 2010 (*Iron Age New Steel*, 1999; Madias, 2014; Stubbles, 2000; Toulouevski & Zinurov, 2010). Electrode diameters increased from the previously common 6.1 cm to 7.1–7.5 cm, and the highest voltages rose from 1000 V to as much as 1600 V. Moreover, all processing required to obtain specific steel qualities was moved out of EAFs to secondary ladle treatment, a shift helping to increase the overall efficiency.

DC furnaces came into operation with the more common availability of reliable sources of that current; the worldwide shift from AC to DC EAFs began in the mid-1980s (Jones, 2003; Takahashi, Hongu, & Honda, 1994). The main advantages of DC furnaces are a lower specific demand for electricity, electrodes and refractories, and reduced operating noise. But these advantages are offset by higher furnace costs and higher specific cost of large-diameter electrodes and have been reduced thanks to the advances in AC operation (Madias, 2014). While the specific power of furnaces with capacities of 50–100 t was between 200 and 250 kVA/t, ultra high-power furnaces (first used in the United States in 1963) have capacities well in excess of 100 t, 70–80 MVA transformers, and specific power input up to 1 MVA/t, with Simetal's Ultimate design reaching 1.5 MVA/t (Siemens VAI, 2012).

Refractory linings are now mostly made of water-cooled panels for longer durability. Use of oxygen has increased productivity and reduced electricity demand. During the 1990s, typical injections were 10–15 m^3/t of steel, and higher rates, now commonly 40–50 m^3 and up to 70 m^3 (when the gas is also used for postcombustion of CO), have been combined with the concurrent injection of carbon (up to 15–17 kg/t). This was necessary to prevent the formation of larger amounts of oxidized iron and the resulting decline in yields. In addition, injected carbon produces foaming slag and arc immersion in the slag improves the efficiency of energy use.

EAFs have remained competitive and gained higher market shares thanks to a continuing quest for higher productivity by cutting operating and maintenance costs. This effort has been particularly important at smaller mini-mills with a single furnace. Relative prices of scrap and hot metal determine the ratios of charged materials and the degree of profitability: lower prices led to increased scrap charging. Common advances have included scrap preheating using oxy-fuel burners, introduction of coal and other carbon additives, and postcombustion of CO produced during the conversion. For furnaces relying on scrap, the raw material amounts to two-thirds of operating cost.

EAFs are now used to produce all kinds of steel, from low-carbon alloy to stainless grades, and they have flexible charging, able to use not only pig iron but also HBI and DRI (Siemens VAI, 2012). They are installed on an upper level above the shop floor (the earlier models were commonly at grade level and required dug-out pits for discharging). The furnace's cylindrical shell sits on a spherically shaped bottom and is covered by a flattened curving roof, its sidewalls above the slag line are covered with water-cooled panels, and the entire interior is lined with refractory material. Electrodes are lowered and raised through the central portion of the roof, and the furnace can be tilted for easy tapping.

Gantry design has a self-supporting roof, oxygen consumption is up to 45 m^3/t, and carbon is both charged (about 10 kg/t) and injected (about 7 kg/t). They have high-power inputs (as much as 1.5 MVA/t) and arc voltage up to 1500V, melting power up to 130MW, and electricity consumption less than 350kWh/t, and those with efficient energy recovery and 100% scrap preheating (to at least 600°C and up to 800°C) require less than 280kWh of electricity per tonne of hot metal. The tap weight of the largest furnaces is in excess of 300t: Simetal's EAF Ultimate in Gebze mill (Çolakoĝlu Metalurji, Turkey) has a capacity of 315t, the world's largest furnace transformer (240MVA), melting power of 205MW, tap-to-tap time of 55min, and electricity requirements of 350kWh/t (Siemens VAI, 2012; Toulouevski & Zinurov, 2010)—but microdesigns (with tap weights less than 35t and annual output of 50,000–200,000t) for small foundries and mills are also available. The other important developments have been the introduction of oxygen blowing (during the 1970s), hot heel, and foaming slag.

Unlike the EAF's complex and expensive equipment (including the electrodes), oxy-fuel burners delivering an equivalent amount of heat are simple and cheap. Oxygen was first introduced through consumable handheld pipes, then by using supersonic lances, and finally through fixed wall injectors obviating any door opening (Madias, 2014). The hot heel practice leaves behind up to 15–20% of hot metal and some slag at the furnace bottom after each tapping; this eliminates the possibility of damaging bottom refractories by electric arcs and allows operating with higher electric power and commencing auxiliary oxygen blowing right after scrap charging. A recent trend has been toward more massive hot heels, even up to 50% of the heat weight (Madias, 2014). Foaming slag, introduced during the 1980s, caused by the ascent of small CO bubbles, is generated by concurrent blowing of oxygen and carbon into the bath; it results in a complete immersion of electrodes, prolongs their life, shields the refractories from the arcs, protects the furnace's roof and walls from excessive heating, speeds up smelting by increasing heat transfer to steel, and reduces electricity requirements.

In the United States, EAF output began to rise slowly, from 5.4Mt of steel in 1950 to 7.6Mt in 1960, and then it jumped to nearly 18Mt in 1970 (Hogan, 1971). EAFs made particular large inroads in the United States with the rise of mini-mills. The shift from BOF to EAF severed the link between primary ironmaking (including coking and ore beneficiation) and steelmaking, and it favored setting up smaller mills that could be located without any regard to the sources of coal, ore, or fluxing materials. The rise and the expansion of mini-mills (plants with annual capacities as small as 50,000t and as large as 600,000t of metal) became the most dynamic and the most competitive component of US steelmaking. Mini-mill output was initially geared toward low-grade products (bars, structural shapes, wire rods), but later many enterprises began to specialize in higher grades of steel (Szekely, 1987). Mini-mill expansion was also helped by their low capital cost (no more than 15–20% of the total needed for integrated steelmaking) and low operating costs when EAFs were combined with continuous casting (Jones et al., 1998).

By 1975, there were more than 200 mini-mills worldwide, with nearly half of them in Western Europe and a quarter in the United States; by 1990 the output of US mini-mills surpassed 33 Mt of steel, and by the year 2000, when one-third of the world's steel originated in EAF, the US steel:pig iron ratio stood at 2.1 as the country's steel output was split between EAFs and BOFs. By 2013, the United States became an even greater consumer of scrap as the steel:pig iron ratio rose to 2.87 and as American EAFs produced just over 60% (52.6 Mt) of all steel, compared to 40% in the EU, 23% in Japan, and less than 9% in China (WSA, 2014). The share of EAFs in global steel production was rising steadily during the last two decades of the twentieth century, and it was expected to surpass 50% by 2010 (Manning & Fruehan, 2001). But the share was still less than 30% by 2013, mainly due to the combination of China's dominance of global steelmaking and the country's very low share of EAF production.

Secondary refining of steel produced by BOFs and EAFs reduces the presence of all undesirable elements below specified levels. Modern mass-produced high-quality steels should contain less than 50 ppm of phosphorus, less than 5 ppm of sulfur, 10 ppm of carbon, and 1.5 ppm of hydrogen (Iwasaki & Matsuo, 2012). Secondary refining (desulfurization, dephosphorization, and decarbonization) is done in converters or in ladle furnaces (Kumakura, 2013), while concentrations of dissolved H_2, N_2, and O_2 are reduced in degassers.

CONTINUOUS CASTING

Post-WW II changes in the treatment of steel after its production have been no less important than was the shift to BOFs and EAFs. Long-standing practice was inherently energy intensive as the hot metal was first cast into steel ingots, oblong pieces that could weigh between 50 t (for specialty steels) and 500 t (for steel destined to be forged into large pieces) and had to be reheated before they could go through a primary rolling mill, where they were formed into one of the three basic kinds of semifinished shapes: wide (some more than 3 m) and thick (up to 25 cm) slabs; square-profile (up to 25 cm) billets; and rectangular-profile blooms (commonly 40 × 60 cm). Final processing (hot- or cold-rolling) turned standard- and medium-thickness slabs into thin slabs and plates, billets into bars and rods, and blooms into I and H beams. The complete casting and rolling sequence could require no less energy than the steelmaking itself, but it persisted for generations, being the final component of the traditional combination that began with blast furnace pig iron and continued with open hearth steelmaking.

Steps eliminated by continuous casting include the pouring of hot metal into ingot molds, removing the molds from the ingots, putting the ingots into soaking pits in order to equalize their temperature, and then rolling them into semifinished products (slabs, billets, or booms). Much like the idea of the oxygen furnace, the concept of continuous casting had also originated with Henry Bessemer: his first patent for a twin-roll caster (using water-cooled rolls with edges sealed by a flange or a dam in groove) was granted in 1865, and the original design was improved in 1891 (Bessemer, 1891), a couple years after R.M. Daelen received a German patent for a vertical casting machine. But Bessemer's peers were unimpressed:

> *One need not, therefore, be greatly surprised that the production of continuous sheets direct from fluid iron did not excite a great amount of enthusiasm in the minds of tin plate manufacturers of that day; in fact, the whole scheme was simply pooh-poohed and laid aside, without any serious consideration of its merits* ***(Bessemer, 1891, p. 27).***

Problems with metal sticking and uneven cooling prevented the conversion of these early designs into functioning machinery, and the first successful commercial applications of continuous casting, during the 1930s, were not for steel but for the castings of color metals with significantly lower melting points. Continuous casting of steel owes it eventual success to the combination of the metallurgical expertise of Siegfried Junghans (1887–1954) and entrepreneurial effort of Irving Rossi (1889–1991). A key to this advance was the invention of a vertically oscillating (reciprocating) mold by Siegfried Junghans: it eliminated the possibility of the cooling metal sticking to the mold.

The first working prototype was ready in 1927; Junghans filed a patent claim in 1933 (inexplicably, it was not granted until 1944) and afterwards devoted himself to adapting the process for steel casting. When, in 1936, he demonstrated his technique with brass casting to Irving Rossi, an American engineer doing business in pre-WW II Germany, it had actually ended in failure as the brass billet skin tore open and hot metal spewed out. But Rossi had correctly distinguished between the challenges with a prototype and the far-reaching commercial potential of the demonstrated technique and immediately secured exclusive rights to Junghans' patent and to its follow-ups for the United States and England and nonexclusive rights for all countries outside of Germany, and in 1938 he added an agreement for sharing information required to build new continuous casting plants in return for financing such developments outside of Germany (Tanner, 1998).

Rossi's first American commercial installation came in 1937 with German-built brass casting, but the need for rapid and massive expansion of wartime steelmaking favored the use of well-established methods, and so it wasn't until 1947 that Rossi persuaded Alleghany Ludlum Steel to embrace the method. The first good-quality slabs (33 × 7.5 cm, and up to 10 m long) were made in May 1949 in the company's Watervliet plant by an American-made (Koppers) casting machine, but this early attempt did not lead to a further commercial adoption. Junghans made his first successful continuous steel casting also in 1949 in his workshop, and in 1950 Mannesmann (the other owner of basic patent rights) acquired the rights and began to build the first German commercial line at Huckingen, and it began operation in 1952. Rossi expanded his promotion and licensing activities by establishing Continuous Metalcast Corporation and, in October 1954, Concast AG in Zurich. Its general manager, Swiss lawyer Heinrich Tanner, eventually wrote a detailed definitive history of the early era of this fundamental technical innovation (Tanner, 1998).

During the next three decades, Concast dominated the global expansion of continuous casting through two highly profitable arrangements. In return for licensing its key continuous casting patents the company first received, gratis, all information and patents arising from the operation of licensed plants. This arrangement was further strengthened in 1970 when Concast made a cooperative patent exchange agreement with Mannesmann, its principal competitor. Second, builders of casting machinery channeled their sales contracts solely through Concast, receiving in return worldwide marketing, and consulting services guaranteed to resolve any technical problems arising during the early phases of commercial operation. As a result, Concast had eventually controlled more than 60% of the global market for continuous casters, it was identified as a virtual monopoly, and in 1981 it had to be reorganized pursuant to antitrust rulings in the United States and Europe.

DIFFUSION AND IMPROVEMENTS

Diffusion of continuous casting was only slightly different from the experience with BOFs: Japan raced ahead, and while the US steel industry was not such a laggard as it was with oxygen furnaces, it moved slower. Although the first Japanese continuous casting line began to work in 1955, most of the expansion took place during the 1970s when the continuous casting share rose from 6% to 60%, and by 1990 more than 90% of Japan's steel was made that way (Okumura, 1994). The first continuous casting plants in the United States went into operation in 1962, at US Steel in South Chicago and Roanoke Steel in Virginia, followed by a fairly rapid adoption during the late 1960s (Hogan, 1971). As a result American continuous casting capacity rose from just 65,000 t in 1962 to 14.4 Mt by 1970, and the years between 1968 and 1977 saw the largest number (more than 1100) of new continuous casting patents, with half of them filed by American inventors, but the US adoption of the technique continued to lag: by 1975 continuous casting reached just 10% of US steel production, it rose above 20% by 1980, and it reached 66% by 1990 (Aylen, 2002; Burwell, 1990).

By the century's end, the technique became universal, claiming nearly 87% of the global steel output, virtually all (96–97%) production in Japan, the EU, and the United States and 87% of the total in China, with only Russia (at 50%) and Ukraine (at a mere 20%) lagging far behind. The earliest models of continuous casting machines had simple vertical designs and hence they required tall structures or deep excavated pits. Designs capable of bending the hot metal strand (first used in 1956) reduced the assembly's height by nearly a third; curved bow-type casters, commercially available since 1963 (and developed independently in Switzerland at von Moos Steelworks and in Germany by Mannesmann), brought further height reduction by at least 40%, and low-head and horizontal designs became common starting in the 1980s (Okumura, 1994).

The standard sequence of modern continuous casting is as follows (Luiten, 2001; Morita & Emi, 2003; Schneider, 2000; Schrewe, 1991; Tanner, 1998). Molten steel (from the oxygen or electric arc furnace) is poured into a ladle that is transported to the inlet of a continuous casting machine, raised onto a turret, and poured into a tundish, a vessel whose large volume ensures the continuing flow of hot metal through a submerged entry nozzle and into a water-cooled and vertical copper mold, where the solidification begins as an outer shell starts enveloping a liquid core. The mold oscillates in order to minimize friction and eliminate any sticking, shell tearing, and liquid steel breakouts that would stop the casting and require expensive cleanup and repairs. As soon as the metal's edge zone solidifies (1–2 cm), the withdrawal unit draws the strand out and the rollers move the partially solidified strand at speeds matching the flow of new supply.

Casting speed depends on the profile and quality of the cast metal, and for standard slabs, billets, and blooms it may be as slow as 30 cm and as fast as 7.5 meters per minute, with the most common rates between 1 and 2.5 meters a minute. Primary cooling takes place in the mold zone; the secondary cooling is done by water sprays as a new slab is drawn forward and gets bent on support rolls without any deformation or cracking. The strand gets straightened, and at the caster's end, between 10 and 40 m from the hot metal inlet, even its core becomes solid and the metal is cut by oxyacetylene torches and discharged to be either stored before further processing or hot-charged for final rolling (Fig. 5.4). The process can be configured to produce a variety of semifinished steel castings. Early designs would cast only square or round billets of limited dimensions. Later machines could handle much larger blooms and slabs and produce near-net-shape profiles.

The most rational step would be to move the continuously cast hot slabs directly to the rolling mill to make finished products, a sequence that would bring further energy savings and reduce the inventories of semifinished profiles, but established plant layouts do not usually allow such

Figure 5.4 Torches cutting continuously cast steel at the Novolipetsk Steel mill in Kaluga, Russia. *Corbis.*

additions, and the high capital costs of retrofitting make it less appealing. More importantly, since the late 1980s it has also been possible to make thin slabs (5–6 cm thick, casting speed of 4–6 meters a minute, to be hot-rolled directly into strips), and a decade later also thin strips (1–5 mm, casting speed 15–120 meters a minute), that can be coiled at the end of the process. Development of thin slab casting started during the 1970s, and Germany's Schloeman Siemag, a leading maker of casting machinery, had the first prototype in 1985. Nucor in Indiana operated the first thin slab caster in 1989, and during the 1990s other designs were introduced in Japan, Italy, Austria, and Sweden.

The greatest challenge was to master direct production of thin strips, an effort that involved a number of secretive, and expensive, development projects in all major steel-producing countries during the 1990s (Luiten, 2001). The process does not use any oscillating molds, as molten steel goes directly between two water-cooled drums, and as it cools about 1000 times faster than during slab casting, it gets compressed by rotating drums into strips just a few millimeters thick (Schneider, 2000; Takeuchi et al., 1994). In a typical bow-type caster used to produce steel strips, temperature falls from 1500 °C to 1150 °C by the time the metal leaves the withdrawal rollers, and rapid rates of cooling (with heat fluxes reaching as much as 20 MW/m^2 and temperature differences of up to 300 °C/cm in the forming strip compared to just 1–5 MW/m^2 in oscillating mold casters) allow high casting speeds of more than 100 meters a minute, but it is a challenge to avoid premature solidification at edges and to produce a uniform strip with the exit temperature of about 650 °C.

Nippon Steel and Mitsubishi Heavy Industries began to operate the first strip-casting line (1.33-m-wide and 2–5-mm-thick stainless strip) in secret in the fall of 1997 and announced its success a year later, when Australia's Broken Hill Proprietary Company also revealed its line at Port Kembla in New South Wales capable of casting strips of 1.5–2.5 mm that were reduced by an inline rolling to 1.1 mm (Bagsarian, 1998). Cooperation of three European companies resulted in the Eurostrip (for austenitic stainless steel, 1.5–4.5 mm thick and 1.1–1.45 m wide, average casting speed 60–100 meters a minute, coil weight 30 t) launched at ThyssenKrupp's Krefeld plant in December 1999 (Bagsarian, 2000).

Continuous casting is now a mature production process, with new casters built and some old ones reconfigured, and with incremental performance improvements aimed at increasing throughput with higher casting speeds and at reducing the incidence of costly breakouts when molten steel flows through a defective part of a solidifying shell and makes costly damage to the casting line that requires extensive repairs. In Japan, these efforts to raise the net working ratio of caster lines have lowered the average incidence of breakouts per caster from 5.5 per year in 1990 to 2 in the year 2000 and to less than 1 incident in 2011 (Yamaguchi, Nakashima, & Sawai, 2013).

Continuous casting offers six major advantages when compared to traditional ingot casting that is followed by primary rolling: the process is an order of magnitude faster (30–60 min compared to 1–2 days for the same mass of processed metal); its metal yield is significantly higher (as much as 99% compared to less than 90%); it saves 50–75% of energy; its labor productivity is also at least 50% and up to 75% higher; capital savings are commonly around 60% compared to the traditional setup, and elimination of expensive hot strip mill by direct strip casting can lower the overall capital expenditure by up to 90%; and it requires much less space (Okumura, 1994; Schneider, 2000).

Not surprisingly, this combination of advantages ensured the technique's rapid worldwide adoption. And continuous casting provided the last link in the modern steelmaking sequence which has been most commonly installed in mini-mills. By 1950, the dominant sequence was from blast furnace to open hearth furnace to cast ingots processed after reheating in a rolling mill; by the year 2000 that sequence might, or might not, include blast furnace, its second step was either BOF or EAF, and the crowning achievement was continuous casting of products ranging from massive slabs to thin coiled strips.

The final step is to turn the cast steel into shapes that will leave the mill. The most important products from continuous casting of slabs are hot-rolled and cold-rolled sheets, coated sheets, and electrical sheets. Slabs put through a plate mill end up as plates or UOE pipes (Fig. 5.5). Billets are turned into H-shapes, sheet piles, reinforcing bars, wire rods, and seamless pipes, and blooms end up as rails and light and heavy sections (Cullen et al., 2012). Some steel products are heat treated: this is done in order to harden the metal (using natural gas to heat them to specific temperature before rapidly cooling them by quenching into oil, water, or brine); to temper it in order to reduce brittleness (letting it cool after heating); and to anneal it in order to make it more ductile (keeping it at a specified temperature for a prescribed period, then cooling it slowly).

Detailed, and reliable, Japanese statistics allow us to follow the partitioning of crude steel as well as all major finished products (JISF, 2015).

Figure 5.5 Slab casting at JFE's Fukuyama Works (Hiroshima province). *Reproduced by permission from JFE Steel.*

Data for 2013 show all but 0.3% of pig iron (used by foundries and ferroalloy production) going for steelmaking. Nearly 80% of all crude steel (about 79% of it coming from BOFs and the rest from EAFs) is sold as ordinary steel; the rest are specialty products. About 89% of steel is hot-rolled, with heavy plates, bars, and shapes being the three most common products made from ordinary steel, and strips, bars, and wire rods dominating hot rolling of specialty steel. The breakdown of all finished steel products is as follows (all shares are rounded): hot-rolled strips 18%, galvanized sheets 13%, bars and plates each 11%, shapes and cold-rolled sheets each 7%, and wire rods 2%.

CHAPTER 6

Materials in Modern Iron and Steel Production

Ores, Coke, Fluxes, Scrap, and Other Inputs

The next logical step—after tracing the history of iron and steel production, and after focusing on the role of technical advances in raising output and productivity, expanding the choice of final products and improving their quality—is to go beyond the confines of the plants where smelting, casting, oxygen blasting, rolling, and finishing take place, as well as looking at the production processes from the perspective of energy costs and environmental consequences. The latter two subjects will be the topics of the next chapter; in this one I will take a closer look at the key material inputs required to produce pig iron and steel.

These inputs are dominated by the requirements of pig iron smelting, by massive (and increasingly intercontinental) deliveries of iron ores (now rarely charged as raw mine products but only after being subjected to specific beneficiation and agglomeration processes), coke, and fluxes (mostly limestone)—but they also include less massive, but no less essential, supplies of powdered coal, cooling water, and refractory materials. As already noted, the maximum daily output of large (>3000 m^3) blast furnaces (BFs) is on the order of 13,000 t of pig iron, a flow that necessitates massive material inputs on an annual basis. Such furnaces require every year more than 4.5 Mt of hot air blast, and their total consumption of agglomerated ores (as sinter or pellets, now rarely as crushed ore), coke, injected coal, and fluxing materials (mostly incorporated in agglomerates) sums up to more than 12 Mt of raw materials, all to be charged, in an essentially punctiform way, through a furnace's top. After appraising these inputs, I will present typical material balances for a number of current commercial BF practices.

Material requirements of basic oxygen furnaces (BOFs) are less complicated than the inputs into primary iron production: dominated by hot pig iron, they also include a significant amount of scrap metal, necessary fluxes, and, of course, a reliable supply of oxygen and cooling water.

Still the Iron Age.
DOI: http://dx.doi.org/10.1016/B978-0-12-804233-5.00006-3

Modern steelmaking has become increasingly dependent on recycled metal, and before I will quantify typical material balances of electric arc furnaces (EAFs), I will present key information on the categories, generation rates, accumulated stocks, recycling rates, and international trade of steel scrap, as well as on the limitations of ferrous recycling.

MATERIALS FOR BFs AND BOFs

BF burden used to consist of crushed and sized iron ore and fluxing minerals (mostly limestone) charged with coke (to support the burden and to act as the reducing agent), all added via mechanical conveyors through a sealed top. Typical modern burden is a sintered or pelletized ferrous charge that incorporates flux materials, and reduced amounts of coke accompanied by direct injections of pulverized coal (the latter charge introduced through tuyères rather than through a sealed top). Detailed American statistics allow us to trace the reduction of specific raw material inputs since the beginning of the twentieth century (Gold et al., 1984; Kelly & Matos, 2014).

For iron ore (and/or its agglomerates), average charge per tonne of pig iron declined from about 2.1 t in 1900 to less than 2 t by 1925 and changed little during the next 25 years; by 1960 it was down to about 1.75 t, in 2000 it was just above 1.5 t, by 2010 it was just below 1.5 t, and in 2013 it averaged 1.35 t (WSA, 2015). Specific coke charging declined steadily from about 1.3 t per tonne of pig iron in 1900 to just below 1 tonne by 1932; little changed during the next two decades, but by 1960 the rate was about 820 kg/t, in 1970 it was reduced to less than 650 kg/t, by 1990 it was about 500 kg/t, and it averaged only about 400 kg/t in the year 2000 and 320 kg/t in 2015.

But, as already explained in the previous chapter, this decline does not reflect the overall decrease in carbon needs as it has been accompanied by increased direct coal injection, whose average rose from less than 100 kg/t during the early 1990s to maxima of about 200 kg/t by 2015. Japanese data (Naito, Takeda, & Matsui, 2015; Takamatsu et al., 2012) provide another illustration of this shift. In 1945, Japanese BFs averaged more than 1500 kg of coke/t of pig iron, by 1950 coke-only smelting needed just over 900 kg of coke, and by 1975 the coke ratio was down to 450 kg/t but commonly used oil injection was adding about 100 kg/t. Following the two rounds of rapid oil price rises, all Japanese BFs became oil-less by 1982 and pulverized coal injection began to rise to reach an average of about 130 kg/t by the century's end, in addition to 370 kg of coke, with a maximum monthly PCI rate of 266 kg/t and minimum coke use of less than 300 kg/t.

The US national mean for charging of fluxing minerals (limestone and dolomite) was around 380 kg/t of pig iron during the earliest decades of the twentieth century; subsequent slow rise brought it to as much as 480 kg/t by 1948, and then the rates declined to 330 kg/t in 1960, and during the last decade the rates were only around 275 kg/t. But the fundamental difference is not in charged rates but in the mode of typical use: with the near-universal adoption of agglomerated (beneficiated) ores, calcined limestone and dolomite are not charged directly as raw materials but overwhelmingly as self-fluxing sintered or pelletized ores.

Iron Ore

Iron ore dominates the material input into BFs, and its crustal abundance means that the production of primary iron has not been, and in the foreseeable future will not be, limited by the availability of this key resource. Iron is the third most abundant mineral element in the Earth's crust, with 5%, following Al at 8.1% and Si at nearly 28%. Its content is between 2% and 3% in sedimentary rocks, 8.5% in basalt and gabbro. Iron is mostly found in oxygenated forms (and O_2, with 46.6%, is the most abundant crustal element); contributions by carbonates, sulfides, and silicates are minor. Magnetite (Fe_3O_4), a dark grey or black oxide that can be found in igneous, metamorphic, and sedimentary rocks, has the highest iron content (72.36%), and its strong magnetism makes it easy to separate it from gangue and to produce a high-quality concentrate. Pure hematite (Fe_2O_3) contains 69.94% Fe; it, too, can be found in all kinds of rocks, and it is concentrated by gravity and flotation techniques. Hydrous oxide—above all goethite ($HFeO_2$) and lepidocrocite ($FeO(OH)$)—has just over 60% Fe, and siderite ($FeCO_3$) is the most important carbonate ore, with 48.02% Fe.

Most iron is found in sedimentary ores belonging to ancient (Precambrian) banded iron formation (BIF), metamorphosed sediments including the minerals itabirite, jaspilite, hematite-quartzite, and taconite; these silica-and-magnetite layers can be hundreds of meters thick and can extend for hundreds or even thousands of kilometers. These ores are commonly present in formations that are readily accessible by surface mining, and many large iron ore mines are among the world's most extensive excavations stripping vegetation from large areas and turning them into landscapes of terraces, pits, and ponds. Ratios of material initially excavated to the final marketed product rise with the quality of shipped ore, and they commonly range from 4:1 to 7:1 for high-grade production. Adriaanse et al. (1997) used the ratio of just 2 tonnes of overburden per tonne of extracted iron ore, but it is difficult to estimate the most likely error associated with this global generalization.

Average iron content of common ores varies from less than 20% to nearly 70%: weighted mean concentration at producing mines is now just short of 60%, with Brazilian and Australian ores above that mean and the Chinese ores having just 30–40% Fe. America's leading iron resource, Mesabi Range taconite in Minnesota and other iron ranges in Michigan, has 30–40% Fe and 40–50% SiO_2, and its development (and large-scale enrichment) started only in the 1950s in an effort to reduce reliance on imported ore (Cannon, 2011; Kakela, 1981). Lake Superior taconite, upgraded in pellet form, now supplies about 85% of the US iron ore demand. Since 1950, typical run-of-mine production has shifted from coarse hematites and deeper fine hematites during the 1950s and 1960s to coarse and fine hematites and rich itabirites in the 1970s, to fine hematites and rich and poor itabirites during the 1980s and 1990s, to widespread depletion of hematites and poorer itabirites since the year 2000 (Mourão, 2011).

USGS puts the global resources of iron at more than 230 Gt contained within more than 800 Gt of crude iron ores and reserves at 87 Gt Fe and 190 Gt of crude ore (USGS, 2014). The reserve totals imply global *R/P* ratios of 61 for crude ore and 27 for its metallic content. Iron ores are found worldwide, but, as with most minerals, massive deposits (as well as high-quality reserves) are unevenly distributed, with Australia (26%), Brazil (18%), and Russia (16%) accounting for 60% of global iron reserves. China, now the world's largest iron ore producer, has only about 8% of all reserves, while the United States has less than 2.5%. Global extraction totals, available since the beginning of the twentieth century, show more than half a dozen distinct periods. The production of iron ore (gross weight) rose from about 90 Mt in 1900 to 177 Mt in 1913, declined and stagnated during WW I and the first half of the 1920s, reached 201 Mt in 1929, and returned to that level only by 1937. WW II peak output of 235 Mt in 1942 was surpassed in 1950, and then steadily rising demand for steel brought the aggregate to 902 Mt by 1975, the level that was surpassed only in 1987.

For the remainder of the century the production remained between 925 and 1070 Mt, but then, driven by China's extraordinary rise in primary iron smelting (causing both rising domestic production and large imports), it grew by nearly 60% in a single decade (in 2010, the output was 2.59 Gt), and then nearly 25% higher, at 3.22 Gt, in 2014 (USGS, 2014). This means that iron ore production is the third most massive extractive enterprise in the world: its aggregate annual output is surpassed only by the total mass of fossil fuels and bulk construction materials.

Production of iron ore has shifted with the regional dominance of iron smelting. Europe (including the USSR) was the leader until the 1970s; by the century's end China was the largest producer, followed by Brazil and Australia; by 2014, China became even more dominant, while Australia produced more than twice as much as Brazil. But comparisons in terms of actual metal content tell a different story: China still leads, but its average of only 30% Fe shrank its 2012 output to 393 Mt Fe, compared to 315 Mt Fe for Australia (whose ores average 62% Fe) and about 258 Mt Fe for Brazil, where the average iron content is 52% (Mourão, 2011; USGS, 2014).

The magnitude of the largest iron ore deposits and economies of scale in their extraction have led to a steady increase in the trade of this raw material (Polinares, 2012). The shares of traded iron ore rose from less than 40% in the early 1970s to 47% in the year 2000 and to 69% in 2013 (WSA, 2015). Two generations ago, Australia was the largest exporter but it accounted for less than 20% of the total, while Canada was almost as large an exporter of iron ore as Brazil, and Sweden and France were still among the top 10. Some 40 countries now export iron ore, but Australia and Brazil dominate the global market: in 2013 they sold 70% of all traded iron ore, and by 2020 they are expected to control 90% of the world's seaborne ore trade (Jamasmie, 2014).

Rio Tinto is Australia's dominant company, producing from 15 mines, moving increasing shares of extracted ores by giant autonomous trucks, and having 1600 km of rail lines linking them with shipping terminals; the company is now developing the country's largest integrated mineral extraction project in Pilbara, whose annual capacity is to reach 360 Mt in 2017 (Harding, 2014). Brazil's production and exports are dominated by good-quality itabirite (50% Fe) in Minas Gerais (nearly 70% of nationwide output), and the rest comes from high-grade hematite ores (60% Fe) in Pará (Fig. 6.1). Vale is by far the largest producer (nearly 85%), followed by CSN, Samarco, and MMX (Brazilian Mining Association, 2015). Vale is also the world's largest iron ore exporting company, followed by Rio Tinto and BHP Billiton. Intercontinental shipping of iron ore is done mostly in Capesize vessels (capacity of 140,000 deadweight tons), but much larger ships have been used to carry Brazilian ore, with Vale acquiring 35 362-m-long Valemax carriers rated at 400,000 deadweight tons (Vale, 2015).

Figure 6.1 Aerial view of Carajas iron ore mine in Pará, Brazil. *Corbis.*

China has been the largest iron ore importer since 2003, when it surpassed Japan: in 2013, it bought 820 Mt, or 64% of the world's iron ore exports, and that total accounted for nearly 70% of the ore used in the country (WSA, 2014). Other large importers are Japan (135 Mt in 2013), South Korea, Germany, Poland, and Taiwan. China's extraordinary growth of iron ore imports was the primary cause for a 15-fold increase in ore price between 2000 (when China imported just 70 Mt) and 2011, when the spot prices peaked at nearly $190/t; this has been followed by a fluctuating decline, with spot prices below $70 at the beginning of 2015, but major producers believe that there will be further increase in demand, particularly in Asia.

Direct charging of crushed natural iron ore (whose lumps are sized at 6–30 mm for BFs and 6–18 mm for DRI (direct reduction of iron) plants) has become less and less common as the iron content of ores has declined and as various beneficiation (concentration), sintering, and pelletizing processes are used to improve the ore quality and enhance its suitability for efficient reduction (Poveromo, 2006). Efficient operation of BFs requires rich iron burden (preferably in excess of 58% Fe) and minimum shares of iron ore fines (particles of less than 5 mm), whose presence would impede gas flow and interfere with the reduction process.

When narrowly defined, beneficiation entails removal of large quantities of nonferrous minerals, mostly SiO_2 (reducing its presence to less than 10% by mass), and it is done by the sequence of milling (crushing, grinding) raw ore, its washing, sorting, sizing, magnetic separation, dewatering, and filtering. Flotation following magnetic separation is often used to upgrade concentrates by reducing their silica content. Broadly defined, beneficiation also includes agglomeration that takes mostly the form of pelletizing or sintering (USEPA, 1994). Sintering is the most common, and also most economical, form of thermal agglomeration.

The process starts with the preparation of a raw mix of iron ore and spillage fines (sinter feed is typically 0.15–6 mm), coke fines or coal, and flux (lime, limestone, dolomite); water is added, and the mix goes into a sintering machine (onto a traveling grate), where it is ignited and the coke in the mixture provides the necessary heat (1300–1480 °C) to melt the surface and agglomerate the mix, forming a porous cake that is cooled and crushed into lumps of 15–25 mm suitable for BF burden (Outotec, 2015a). Typical material input is 2.5 t of raw mixture to produce a tonne of sinter. The process was developed to treat fine waste in the early twentieth century and sintered ores are now the dominant material charged into BFs, used to produce about 70% of all hot metal.

Adding basic flux to sinter (that is, producing self-fluxing material) improves its physical and metallurgical properties (it acts as a binder and improves fine particle agglomeration), reduces the melting temperature of the ore, and eliminates direct charging of limestone or dolomite. Typical sintered ores contain about 57–58% Fe, 7–8% CaO, up to 2% of MgO, and 4–5% SiO_2. Because sinter is prone to crushing during transportation, its production is usually done on ironmaking sites from ores shipped from domestic sources or brought by bulk carriers to seaside iron and steel mills (but care must be taken when shipping iron ore as well because its fines can liquefy in a ship's hold). Sintered ore may be the only burden, but more often the charge is composed of a mixture of sintered and pelletized ore, sometimes with a small amount of raw ore.

In contrast to sinter, which is porous and brittle, iron pellets, produced by agglomerating very fine-grained iron ore with binders (bentonite, small amount of limestone), are compact and hard and can be transported over long distances without crushing. Pellets (spheres of 9–16 mm in diameter) are formed from wetted ground ore in balling drums. Then they go into indurating furnaces, moving on a traveling grate on a 30–55-cm-thick bed to be heated (usually by natural gas burners) and then cooled and stored for shipment (Outotec, 2015b). Their thermal treatment (induration) produces a strong and fairly uniform charge that contains more than 60–65% Fe, 4–5% silica, and traces of Al, Mn, and P. Pelletization is done usually right at the mining sites or at the exporting ports, and processing units have capacities of up to 7.25 Mt/year. Global production of beneficiated ore rose from just over 900 Mt in 1990 to 2.33 Gt in 2013 (Schmöle, Lüngen, & Noldin, 2014).

Metallurgical Coke

Decline in specific coke requirements has been accompanied by substantial expansion of pig iron smelting and hence by absolute increase of coal extraction destined for ironmaking. Coking coal should have low ash and low sulfur content, and coals of different volatilities are blended to control the quality of the final product and the volume of by-product gases (Díez, Alvarez, & Barriocanal, 2002; Mussatti, 1998). As coal was displaced in railroad and water-borne transportation by liquid fuels, and as the main source of household, commercial, and industrial heating by fuel oil and natural gas, it retained its worldwide importance in only two major markets: in centralized electricity generation and as the source of metallurgical coke.

Worldwide extraction of coking coal was just above 1 Gt in 2013, with China being by far the world's largest producer with 527 Mt (Fig. 6.2). Australia, with 158 Mt, was a distant second, and expansion of ironmaking in Asia led to rising exports of coking coal: Australia (154 Mt in 2013), the United States, and Canada have been the largest sellers, while China (77 Mt in 2013), Japan, and India have been the largest buyers (WCA, 2015). Coal, delivered by trains and barges or imported in bulk carriers to coastal mills, is first pulverized (to just 0.15–3 mm), blended, and charged from special cars moving above empty hot ovens. Thermal distillation of coal (carbonization at about 1100 °C in oxygen-deficient atmosphere) proceeds in sealed coking ovens with heat transferred from hot brick walls (Valia, 2014; Fig. 6.3).

Coal decomposes at temperatures below 475 °C as it forms plastic layers, first near hot brick walls and then moving toward an oven's center; higher temperatures bring releases of tar and aromatic hydrocarbons and coke shrinks and stabilizes at temperatures between 600 °C and 1100 °C. The coking process lasts 15–18 h to make BF coke and is longer (25–30 h) for foundry coke. Finished incandescent coke is pushed through open battery doors and rapidly quenched (in wet or dry process), crushed, and screened, and it is then ready for metallurgical use. Average lump size is

Figure 6.2 Piles of coal ready for coking at an iron and steel mill in Hefei (Anhui province), in China. *Corbis.*

Figure 6.3 Coke ovens at JFE's Fukuyama Works (Hiroshima province). *Reproduced by permission from JFE Steel.*

about 5 cm across; European and US coke have less than 10% of ash (but up to 13.5% in China), less than 3% of moisture, and less than 1% of sulfur. One tonne of coking coal yields 600–800 kg of BF coke, 50–100 kg of coke breeze (undersize screenings smaller than 1 cm), roughly 300–350 m^3 of coke oven gas, 30–45 L of tar, 10–13 L of ammonium sulfate, and 50–130 L of ammonia liquor (Sundholm et al., 1999).

Mining of metallurgical coal accounted for about 12% of global coal extraction in 2013 (WCA, 2014), and, as repeatedly noted, decline in specific coke use has been accompanied by rising volume of pulverized coal directly injected into BFs. Global production of metallurgical coke rose from about 344 Mt in 2000 to 596 Mt a decade later, and it reached 685 Mt by 2013, while China's rising dominance in ironmaking has raised the country's share of the overall output to 35% in the year 2000, to 65% a decade later, and to 70% in 2013 (Jones, 2014). During the first decade of the twenty-first century aggregate coke output declined in Europe (by more than 20%) and North America (by a similar margin), increased slightly in Latin America, and, thanks mostly to China, expanded 2.3 times in Asia, with the global production rising from nearly 345 Mt in the year 2000 to more than 590 Mt in 2010 and to 685 Mt in 2013.

Fluxing (slag-making) materials are the easiest ones to secure: they come from two abundant and widely distributed crustal minerals, from limestone ($CaCO_3$) and dolomite, $CaMg(CO_3)_2$. In 2012, the United States had nearly 2000 limestone and 150 dolomite quarries (compared to only about 400 granite quarries). USGS provides detailed data for limestone use in construction but withholds data on flux stone in order not to disclose any proprietary information (USGS, 2014). In any case, the two minerals are in abundant supply, and their calcining releases CO_2 and yields CaO and MgO, the two dominant metallurgical fluxes that used to be charged directly with lump ore but now are overwhelmingly incorporated into self-fluxing sintered or pelletized ores. Fluxes combine with SiO_2 present in the ore and absorb sulfur and phosphorus, two undesirable ingredients of pig iron.

Hot pig iron is, of course, the dominant input into BOFs: it can make up to 90% of the total charge, with much smaller additions of steel scrap (the largest heat sink during BOF operation) and fluxing materials as well as small amounts of ferroalloys. Fluxing materials should contain about six times more CaO than SiO_2, and there should be enough MgO to saturate the slag and hence reduce the chemical erosion of the furnace's MgO lining. The lining's longevity is enhanced (to more than 20,000 heats per campaign) by slag splashing after tapping, that is, by blowing the residual liquid slag with pressurized nitrogen introduced through the oxygen lance. Ferroalloys, including manganese, silicon, aluminum boron, and titanium, are added from overhead bins to the ladle.

MATERIAL BALANCES OF INTEGRATED STEELMAKING

The latest estimates by the World Steel Association for typical material inputs required to make a tonne of crude steel by the integrated steelmaking route (BF and BOF) are as follows: 1400 kg of iron ore (mostly as sinter or pellets or their combinations), 800 kg of coal (most of it converted to coke), 300 kg of fluxing materials (mostly for BF), and 120 kg of recycled steel. For a large BF producing 10,000 t/day of iron and supplying an adjacent large BOF, annual requirements add up to 5.11 Mt of ore, 2.92 Mt of coal, 1.09 Mt of flux materials, and nearly 0.5 Mt of steel scrap. A large integrated steel mill thus receives every year more than 9.5 Mt of materials that must be eventually introduced in a virtually punctiform fashion through the sealed top of a BF and into the open top of a BOF.

In global terms in 2012 (with roughly 1170 Mt of pig iron and 1100 Mt of crude steel produced in integrated mills by BF–BOF sequence), this added up to about 1.64 Gt of iron ore, nearly 900 Mt of coal, about 330 Mt of fluxing materials, and 130 Mt of steel scrap. Extracting, producing, transporting, and handling this mass of materials requires a vast global system of mining, shipping, processing, and storage that has to operate without interruption. Inevitably, these massive inputs are transformed not only into the desired valuable metal (crude steel ready for further processing into specific alloys and semifinished and finished products) but also into massive solid or voluminous gaseous waste streams that must be either captured and reused or collected and recycled.

Material balances should also include hot air blast delivered through tuyères placed around the hearth's perimeter (Fig. 6.4). Depending on the operating pressure and on the degree of oxygen enrichment, the mass of hot blast air is mostly between 1.25 and 1.5 t/t of hot metal, and oxygen enrichment ranges between 30 and 300 kg/t of pig iron. Besides the hot metal (pig iron), the output of the smelting process consists of slag and gases. BF is tapped by drilling 2–4 tapholes (the clay-filled openings at the base of furnace hearth, resealed by clay after casting) 8–14 times a day to remove molten iron and slag.

Figure 6.4 Tuyères of Baosteel's (Shanghai) No. 4 BF on its completion day in 2005. *Corbis.*

Because of their very different densities (slag at 2.5 t/m^3, pig iron at 7.2 t/m^3), these two materials do not mix, and the slag, forming a supernatant on hot iron, is easily separated and channeled into a slag pit while iron is transported, in special torpedo cars or in massive ladles, to casting machines. Specific slag quantities vary depending on the amount of charged slagging materials, quality of ores, and efficiency of the smelting process, ranging mostly between 250 and 300 kg/t of hot metal, just a third or no more than half of the typical yield during the early 1950s (Lüngen, 2013; Schmöle, Lüngen, & Noldin, 2014). I will deal with slag's postmill fate in the next chapter summarizing the environmental impacts of iron and steel industry.

Gases generated by the reduction of iron ores and production of hot pig iron are typically composed of up to 23% each of CO and CO_2, and a few percent of H_2, with nearly all of the rest being N_2. Their density is thus about 1.4 kg/m^3, and with volumes ranging mostly between 1150 and 1550 m^3/t of hot metal, the material flow amounts to at least 1.6–2.2 t/t of pig iron. Their energy density is only about a tenth of the energy density of natural gas (mostly between 3.2 and 3.6 MJ/m^3), but that is high enough to justify their capture and reuse. Reusing the top gas requires first the removal of the entrained dust, thus eliminating one of the most polluting by-products of formerly uncontrolled ironmaking.

Material flows in the primary iron smelting in BFs are obviously restricted by many stoichiometrical and physical requirements, but the fundamental identity of the process allows for some significant departures in terms of charge inputs. Coke always dominates the supply of carbon, but it can be partially substituted by powdered coal or by injections of liquid or gaseous hydrocarbons; iron can be charged as a raw ore, as pellets, or as sinter, or as a combination of any of two or three of these materials; and BFs can operate without any supplementary oxygen or with substantial enrichment. Differences in these material balances are easily appreciated by tabulating a few examples of specific material inputs.

The first comparison (Table 6.1) illustrates the range of BF charges attributable to differences in the form and quality of charged ores (lump, sinter, pellets, mixtures of two or all of these), furnace sizes, and productivities, their use of coal injection, and blast's enrichment by oxygen. The table lists actual charges for specific BFs or typical rates used in model calculations, all normalized per tonne of hot metal: they include an example of a typical Belgian–Dutch–German operation (Danloy et al., 2008), a Swedish BF in Luleå (Ryman et al., 2004), a large Russian furnace, Severstal No. 5, and a large Ukrainian furnace, Mittal Steel No. 9 (Tovarovskiy, 2013), Czech (Besta et al., 2012) and Polish (Burchart-Korol, 2013) operations, and typical performances for the EU (IEA, 2010; Remus et al., 2013) and Japan (Morita & Emi, 2003; Nogami et al., 2006).

The second comparison lists material inputs and outputs of three large BOFs in the United States, Japan, and Spain (Table 6.2). Material flows for all natural gas-based DRI processes are similarly simple: 1.7 t of pellets and 0.18–0.24 t of CH_4 (and 100–135 kWh of electricity) per tonne of hot iron, while coal-based processes charge 0.42 t of coke and 6 kg of lime (IEA, 2010).

Naturally, charged materials claim most of the attention in analyzing the balances of BF ironmaking and BOF steelmaking, but water is also an indispensable input. Three uses claim most of the supply: material conditioning (dust control in sintering, slag quenching in BFs, scale removal in hot rolling) consumes about 12% of the total demand; air pollution control (in wet scrubbers to remove dust and gases) accounts for roughly the same share; and heat transfer (mainly for protecting equipment through

Table 6.1 Material Flows in Large BFs

Blast furnaces	Iron ore	Sinter	Pellets	Coke	Coal	Oxygen	Slag
Belgium[a]	1584			209	180	300	260
Belgium[b]	1581			286	197	51	263
Netherlands[c]		1580		300	200	320	240
Sweden[d]			1367	327		56	166
Czech Rep.[e]	104	1283	247	504			396
Poland[f]		1307	250	428			303
EU[g]		1340	155	320	200	50	250
EU[h]	180	1088	358	359			150–347
US[i]	1628				480		255
US[j]	1524			486			220
Ukraine[k]		1345	295	474		102	367
Russia[l]		50	1023	489	412	87	300
Japan[m]	280	1160	190	380	120	40	300
Japan[n]		1317	356	446	57	25	287
Japan TGR[o]		1320	357	357		270	251

(All values are in kg/t of hot metal.)
All values are in kg/t of steel, except oxygen, which is in m^3/t.
[a]Danloy et al. (2008).
[b]Lüngen (2013) (ArcelorMittal Gent A).
[c]Geerdes, Toxopeus, and van der Vliet (2009).
[d]Wang et al. (2008).
[e]Besta et al. (2012); all coke.
[f]Burchart-Korol (2013).
[g]IEA (2010).
[h]Remus et al. (2013), weighted EU average.
[i]Burgo (1999).
[j]Morris, Geiger, and Fine (2011); iron ore includes 47 kg of BF slag with 20% FeO.
[k]Tovarovskiy (2013).
[l]Tovarovskiy (2013).
[m]Morita and Emi (2003).
[n]Morita and Emi (2003).
[o]Nogami et al. (2006); TGR= top gas recycling.

Table 6.2 Material Flows in Large BOFs

Basic oxygen furnaces	Hot metal	Scrap	Ore	Fluxes	Oxygen	Slag	Gases
US (Miller et al., 1998)	877.6	201.6	16.8	56.5	75.3	100.1	100.1
EU (Remus et al., 2013)	860	220	9.7	48.5	49.5–70	125.0	91.0
Japan (Morita and Emi, 2003)	1033.0	28.0	11.0	31.0	66.6	50.0	124.5

(All values are in kg/t of crude steel.)

extensive water cooling) claims about 75% of the total volume (USDOE, 2013). Specific rate per tonne of crude steel ranges widely among, and within, countries, with a world survey indicating extremes of 1–148 m^3/t for input and 1–145 m^3/t for discharge. This means that order of magnitude differences are not uncommon even within advanced countries: Germany's average water consumption in steelmaking declined from more than 35 m^3/t in the early 1980s to 10 m^3/t in 2010, while the country's major steel producer averages only 0.77 m^3/t of crude steel (Lech Stahlwerke, 2011).

Moreover, care must be taken to compare identical volumes, that is, water actually used by a specific process, water that is repeatedly recycled, and freshwater additions. Gao et al. (2011) provide details for all of these categories for the recent Chinese steelmaking practice and the three rates (in the given order and all in m^3/t) are 0.45, 15.47, and 15.97 for coking, 1.75, 68.60, and 70.57 for ironmaking, 0.57, 15.15, and 15.86 for steelmaking, 0.54, 11.86, and 12.40 for continuous casting, and 0.44, 54.70, and 55.63 for hot rolling. These rates (including small volumes for sintering) add up to about 3.8 m^3 of consumed water, 167 m^3 of recycled water, and 172 m^3 of freshly supplied water, with ironmaking accounting for about 40% and hot rolling for more than 30% of freshwater additions. They also put the Chinese national average of consumptive use at 7–8.3 m^3/t, with major enterprises averaging 2.5–6.7 m^3, the highest total being, unfortunately, for the Baotou steelworks in the arid north.

The best recent appraisal of average American requirements puts them on the order of 300 m^3 (300 t) per tonne of steel, a total that includes new supply as well as water that has been recycled and reused (Worrell et al., 2010). Recycling is practiced widely but inevitable evaporation requires between 50 and 100 m^3/t of fresh, make-up water that must be supplied either from public sources or from nearby water bodies. But when Horie et al. (2011) compared water use in steelmaking in Japan, China, and the United States, he used an impossibly low average water footprint (direct and indirect withdrawals) of just 0.62 m^3/t of BF/BOF steel in Japan, 0.99 m^3/t in China, and about 5 m^3 in the United States. In contrast, the latest Polish figures appear quite realistic: direct freshwater use per tonne of crude steel in integrated mills at about 105 m^3, with circulating water amounting to about 35 m^3 (Burchart-Korol & Kruczek, 2015).

Other published specific rates (all in m^3 or t/t of steel) are 5 for sinter plants, 9–10 for EAFs, 12–13 for cold rolling, 13 for continuous casting, 16 for BF, 17 for BOF, 32–40 for various hot-rolling processes, and

37 for coking. Given these substantial differences, quoting a representative global average may be rather imprudent, but the World Steel Association has offered one, based on data collected from 29 steel plants worldwide, and ended up with an average consumption of 28.6 m^3/t and discharge of 25.3 m^3/t for the integrated steelmaking, and, respectively, 28.1 and 26.5 m^3/t for EAFs (WSA, 2011a).

STEEL SCRAP

Steel does not lose any of its many desirable physical properties by being remelted and recast, and hence it can be recycled (inevitable handling and processing losses aside) *ad infinitum*. If care is taken to eliminate undesirable elements from steel scrap, then the recycled material can be used to produce casting components and finished products (be they boilers or bolts) of the same quality as those made from primary pig iron—and do so with a fraction of energy required for primary metal production. As a result steel is the world's most recycled metal. The mass of collected steel is also much larger than for any other recycled material. In the US mass aggregates in 2015 were, with steel equal to 100, about 67 for paper, 6 for aluminum, 4 for glass, and just 3 for plastics (SRI, 2015).

Metal recycling has increased substantially since the 1960s, but the effort still has a long way to go: a UNEP report indicates that only 18 out of 60 metals have end-of-life recycling rates in excess of 50% (fortunately, the list includes not only iron and aluminum but also copper, nickel, manganese, cobalt, and chromium as well the three precious metals, silver, gold, and platinum) and only three rates at between 25% and 50% (UNEP, 2011). Depending on the country, actual end-of-life recycling rates are up to 90% for steel, up to 70% for aluminum, and 60% for nickel but only a bit above 50% for copper.

EAFs are the dominant processor of scrapped ferrous metal that is also charged, in smaller specific amounts, to BOFs. Steel scrap is usually classified into three major streams: home, prompt, and obsolete scrap. Home, or circulating, scrap (also called own arisings) is produced within iron and steel plants and is obviously the easiest material to be recycled, often in a matter of days or weeks after its generation; moreover, its composition is perfectly known, eliminating any quality concerns, and in 2012 it amounted globally to 200 Mt or about 35% of all scrap and the equivalent of 13% of that year's total steel production (BIR, 2014). The rest of the recycled scrap, globally about 65% of the total in 2012 (370 Mt), is purchased domestically or it is imported, and most of it (about two-thirds) is old metal (obsolete scrap).

Prompt (new) scrap originates from a wide variety of metalworking and manufacturing enterprises that use steel to make metal parts, assemblies, or finished products ranging from white goods (refrigerators, washing machines) to heavy land-moving machinery; composition of this scrap is also well known, but it must be collected and transported to mills and hence it becomes available in a matter of months. Estimates of prompt scrap as a share of finished steel products range between 10% and 15% on the all-industrial basis, with rates as low as 8% in the production of containers and as high as 20% for making some car parts (Kozawa & Tsukihashi, 2011). Obsolete scrap comes from old, discarded metal, some of it years and some of it decades old, with steel often containing undesirable elements in alloys and coatings or commingled with other metals and nonmetallic materials in still-assembled appliances, tools, transportation equipment (ships, railway cars, vehicles), and heavy machinery or embedded in waste from demolished buildings.

Ranges of lifespans of steel-based products differ among countries, but the following numbers are good average (and extreme) approximations (Dahlström et al., 2004; Kozawa & Tsukihashi, 2011; Pauliuk et al., 2013): 60–75 (20–100) years for structural steel in buildings, dams, and bridges; 60 years for oceangoing vessels; 30 (10–40) years for industrial machinery (including turbines and engines); 15 (5–20) years for automobiles and durable metal goods; 10 years for boilers and drums; and as little as 1 or 2 years for cans and metal boxes and 5 years for containers. Rail track lifespan is more complicated: before the metal gets recycled, rails can first be reused by swapping the tracks, and later they can be installed on secondary lines with less traffic.

Because the metal is magnetic, recycled steel can be easily separated from nonmagnetic materials, usually in rotating drums after shredding. Screening, pressurized air, and liquid flotation are also used for separation. Admixture of other metals, nonmetallic wastes, and hazardous compounds is a common occurrence with obsolete scrap, whose rusting surfaces can be heavily contaminated by toxic compounds, while the metal itself may contain impermissibly high levels of copper, tin, nickel, chromium, or molybdenum: their concentrations should be generally less than 0.1% for structural steel. Because these metals cannot be separated from the steel, their presence must be diluted by mixing in more high-purity scrap or pig iron. Sized and sorted metal is compacted into small blocks for convenient handling and for land or water-borne transport to domestic steel mills or for international, and increasingly intercontinental, exports.

But in many cases it requires a great deal of effort to get sized, sorted, and compacted blocks of metal ready to be charged into EAFs and also BOFs. Because discarded steel objects come in so many sizes, the recycled material must be first reduced to dimensions suitable for easy handling. The greatest challenge is cutting heavy steel plates of ship hulls and dismantling giant oil tankers, bulk carriers, and cruise liners: some 90% of this difficult work is done (after stripping away the nonmetallic components) by gas and plasma torches at sprawling ship-breaking sites on the seashores and in the shallow waters of Pakistan (Gadani near Karachi), India's Gujarat (Alang), and Bangladesh (near Chittagong). Despite the Hong Kong International Convention for the Safe and Environmentally Sound Recycling of Ships adopted in 2009 (IMO, 2015), thousands of laborers employed at those sites work commonly without any protective gear, and often barefoot, as they perform series of hazardous tasks while walking (and living) on grossly polluted beaches (Rousmaniere & Raj, 2007).

In contrast, car recycling, the dominant source of obsolete steel in affluent Western countries and in Japan, is highly mechanized. After removing some components that could be resold or recycled (stereo, alternator, some engines), and separating larger nonmetallic parts (some of which, including batteries, tires, and plastics, are also partially or largely reusable) and draining recyclable fluids (gasoline, oil, coolant, transmission, and windshield), car bodies are compacted in a crusher by hydraulic machinery operating under high pressure; the flattened bodies are loaded on trucks or railway cars for transport to mills or for export or are shredded at collection sites (Fig. 6.5). In the United States, a country with a high rate and a long history of car ownership, vehicle recycling is a major industry. Annual totals of scrapped vehicles rose from about 2.5 million units in 1950 to 8.3 million vehicles in 1970, reached a peak of 14.3 million in the year 2000, and has since fluctuated between 11.4 and 14.2 million units (USDOT, 2015).

Two countervailing trends affected the total scrapped mass: vehicles got heavier, but the shares of iron and steel declined as aluminum and plastics (and most recently also composite materials) replaced many parts formerly made of cast iron and carbon steel. During the 1950s, ferrous metals made up 94–97% of all metallic components, while their recent shares are down to about 80% of all metals and to between 60% (for hybrids) and 70% for all materials (Field et al., 1994; Jody et al., 2009). On the other hand, cars

Figure 6.5 Flattened cars ready to be shredded and charged into EAFs to make new steel. *Corbis.*

now contain more high-strength steel: its mass is now more than double that used during the 1970s (for more on this aspect see the next chapter on steel uses).

The United States now has about 12,500 vehicle dismantlers, about 2500 scrap processors, and more than 300 heavy car shredders; the industry employs about 30,000 people, and in 2012 it recycled 14.8 Mt of iron and steel from 11.8 million end-of-life vehicles whose mass is now about 60% iron and steel (Fenton, 2014). In 2014, this automotive scrap trade (including exports) was valued at more than $26 billion, and the total made the US automotive recycling the country's 16th largest industry. In 2012, about 88% of domestic scrap consumption was destined for iron and steel mills (of which about 80% went to EAFs and 15% was charged to BOFs), and the ferrous castings industry bought most of the rest.

The recycling rate for automotive scrap fluctuates with ups and downs of the car market: in recent years it peaked at 121% in 2009, and it was only 85% in 2013, with the 10-year (2004–2013) average of 103% (SRI, 2014). The recycling rate of US structural steel (beams and plates) has been nearly as high (averaging 98% in recent years); the rate for obsolete appliance scrap consisting mainly of discarded refrigerators, washing machines, dryers, dishwashers, and stoves and containing about 60% of steel, is about 90%. And despite the fact that reinforcement bars in concrete are obviously much more difficult to recycle, their reuse reached also 70% in 2012 compared to just 40% during the late 1990s. Another major scrap stream comes from containers, mostly from steel cans used by many industries for packaging and distributing food, pet food, aerosols, personal care and home care products, lubricants, and paints.

Steel cans dominated the beverage market until the 1960s, but in the United States they were entirely displaced by lighter aluminum—for beer by 1994 and for soft drinks by 1996. However, in food industry they remain ubiquitous for preserving meats, fruits, and vegetables: in the United States more than 1500 food items are sold in cans (CMI, 2013). Replacing freezing by canning has a number of environmental gains, above all substantial reduction of acidification, air pollution, and emissions of greenhouse gases. As the cans got more common (in the United States more than 100 million of them are opened every day), they got lighter: in the EU the mass of beverage steel cans was halved in 30 years. The cans come as traditional tinplate steel (with a thin tin coating on both sides), tin-free sheet (Cr_2O_3-plated), and polymer-coated steel. In the United States, their recycling rate is now about 70% or close to 2 Mt/year; in the EU, their recycling rate rose from just around 25% in 1990 to the US level, in Japan the rate is 85%, and in China it is about 75% (WSA, 2011b).

In the United States, the overall rate of steel recycling rose from the mean of 67% during the first half of the 1990s to as much as 103% in 2008, and it averaged 90% during the years 2009–2014 (SRI, 2014). In the United States, scrap is destined primarily for two categories of steel mills: the first one makes light flat-rolled products that contain about 30% recycled metal; the other one makes a wide range of products, but it is the exclusive producer of structural shapes whose recycled content is about 80% (AISI, 2009). Estimates of global recycling rates by the World Steel Association are, as expected, somewhat lower than the US values: 85% for construction and automotive, 69% for containers, and 50% for appliances, with the overall mean of 83% in 2007 and target of 90% in 2050 (WSA, 2009). These rates make it clear that, overall, steel recycling has been a great success, resulting in substantial reduction of waste disposal and concurrent energy savings.

Rising steel demand and accumulating stores of steel have resulted in generally higher rates of recycling, expanded international trade (with some peculiar, although not really surprising, attributes), and rising (but also rather unstable) scrap prices. In comparison, aluminum is the next most recycled metal, with 67% of all cans (the most common objects made from this light metal) now recycled in the United States, while no more than about 30% of all consumer electronics, 25% of all glass containers, 10% of all plastics, and less than 10% of synthetic carpets are recycled (USEPA, 2014).

Given the highly uneven distribution of steel stocks, the international steel scrap trade was destined to grow: global exports tripled between 1975 and 2000 (from less than 20 Mt in 1975 to about 68 Mt at the century's end), and they reached the record level of 108.7 Mt in 2011 before falling back to about 95 Mt in 2013 (BIR, 2014; WSA, 1982; WSA, 2015). The United States has been the leading (but fluctuating) exporter, accounting for about 40% of the total in 1975, just below 10% in 2000, 27% in 2010, roughly 30% in 2011, and down to 20% in 2013. In 2014, American steel scrap exports to its main customers declined substantially, by about 20% to Turkey and South Korea, and by two-thirds to China (Recycling International, 2014).

Although some European countries have been steady scrap exporters (Germany, France, the Netherlands) and the EU as a whole is the net scrap exporter, the EU is also the world's largest importer (it bought nearly a third of the traded total in 2013), while Turkey leads the national import ranking with nearly 20 Mt in 2013 compared to 9.3 Mt to South Korea, 5.6 Mt to India, and 4.5 Mt to China. Besides the United States and EU, the other large scrap exporters are Japan (to South Korea and China), Russia (mainly to Turkey and South Korea), and Canada (mostly to the United States). Chinese scrap imports have been fluctuating: they rose from just 1.2 Mt in 1995 to 10.1 Mt in 2005, 2 years later they fell to 3.4 Mt, they peaked at 13.7 Mt in 2009, and they fell again to 4.5 Mt by 2013. Their only constant has been their dominant country of origin, the United States. This aspect of the Sino-American trade is certainly among the most remarkable indicators of shifting national fortunes, with the world's largest affluent economy being the leading supplier of waste materials to the world's fastest modernizing country.

Not surprisingly, scrap collection, sorting, selling, and exporting are highly decentralized, and the world's largest metal recycler, Britain's Sims Metal Management, is processing annually only more than 17 Mt of material, mostly steel. Steel scrap prices grew slowly during the first 7 years of the twentieth century and then more than doubled in the early months of 2008 before plunging by about 70% before the year's end. During 2012, they were as low as $330 and as high as $425; by the spring of 2015 they fell to less than $250 (Steel Benchmarker, 2015).

MATERIAL BALANCES OF EAFs

World Steel Association puts the recent average of specific material requirements for this, the second most important steelmaking route, at about 880 kg of scrap metal, 150 kg of coal, and 43 kg of limestone per tonne of steel (WSA, 2011a). Remus et al. (2013) offers these ranges for EAFs in the EU (all per tonne of steel): 1039–1232 kg of scrap, 0–53 kg of pig iron, 0–215 kg of DRI, 25–149 kg of lime and dolomite, and 3–28 kg of coal. In 2012, when the worldwide output of EAFs reached 446 Mt (nearly 29% of the total), that translated to about 390 Mt of charged scrap metal, 67 Mt of coal, and 10 Mt of limestone—and every tonne of steel made from scrap instead from pig iron saves about 1100 kg of iron ore, 640 kg of coal, and 2.9 MWh of energy. As already noted, the third route, the DRI ores, has not diffused as rapidly as once hoped. MIDREX DRI, commercially the most successful version, requires 1.7 t of pellets, 240 kg on natural gas, and about 135 kWh of electricity (IEA, 2010).

As with the BOF steelmaking, actual material requirements of EAFs vary depending on the kind and quality of the charged metal, and the charges can be fairly flexible, consisting of combinations of scrap with hot metal (pig iron, making up anywhere between 10% and 50% of the total ferrous charge) and scrap with briquettes. For example, material requirements of a typical design of Siemens' SIMETAL EAF Ultimate—with tap weight of 120 t, tap-to-tap time of 30 min, and daily productivity of 5760 t in 48 taps—are 45 m^3 (about 60 kg)/t of oxygen, 10 kg/t of carbon charged and another 7 kg C/t injected, and 1.2 kg/t of electrodes (Siemens VAI, 2011).

In contrast, charges into a large Turkish EAF show the following specific rates (also per tonne of steel): scrap 908 kg; pig iron 136 kg; coke 18 kg; fluxes 45 kg; ferroalloys 6.3 kg; electrodes 2.3 kg; oxygen 63.8 kg; and 10.9 t of cooling water (Yetisken, Camdali, & Ekmekci, 2013). Specific charges into a small (30-t) Turkish EAF were 1053 of scrap, 49 kg of coke, 36 kg of fluxes, and 92 kg of oxygen, while the waste streams included 86 kg of slag, 120 kg of gases, and 19 kg of dust per tonne of steel (Tunc, Camdali, & Arasil, 2012). And for a German furnace using only scrap, Pfeifer and Kirschen (2002) reported average specific inputs of 1036 kg of metal, 21 kg of coal, 4 kg of natural gas, 28 kg of fluxes, 56 kg of oxygen, and 3 kg of electrodes per tonne of steel, and outputs of 16 kg/t of dust and 235 kg/t of furnace gases.

Although EAF is an inferior hot metal processor compared to BOF, charging of pig iron to replace a portion of scrap reduces electricity consumption and raises productivity, and it is now often done in integrated mills where such metal is readily available. The mass of hot (1150–1350 °C) metal poured into the furnace contains energy equal to about 450 kWh/t and, consequently, if it amounts to 40% of the total charge it will reduce the overall electricity consumption by 180 kWh/t (Toulouevski & Zinurov, 2010). But scrap recycling remains the EAF's primary mission, and changes in material handling have contributed to their higher productivity and reduced electricity demand.

Large furnaces used to be charged through the open top from two or three scrap-carrying baskets, prolonging the charging time and allowing the escape of furnace gases, but recent practice has converged on charging even the furnaces with more than 100 t capacity from a single basket (loaded with scrap whose density is up to 0.8 t/m^3), a practice that requires switching off the current for as little as 3 min, and that also halves the dust emissions compared to the charging with two baskets. But larger furnaces, such as the 320-t Simetal Ultimate installed since 2007 at Gebze mini-mill in Turkey (processing scrap and pig iron), use two or three baskets (Siemens VAI, 2012).

Another important energy-reducing measure has been the preheating of the charge scrap. This was done first simply by heating it inside the charging baskets, or in special baskets designed to withstand high temperatures and also cooled by air or water. Two better options use the conveyor method of heating, either in a vertical arrangement (as sectional shaft preheaters) or as horizontal preheaters, with both using the exiting furnace gas. The horizontal belt conveyor, used since the late 1980s by the Consteel process, achieves virtually continuous charging through a sidewall door. Scrap moves through a tunnel (about 30 m long) and is heated by furnace gases and discharged into the furnace at a level that maintains a constant temperature of at least 1580 °C.

Besides increased energy efficiency, this method also reduces arc-generated noise and prolongs the life of water-cooled panels, and because the furnace operates at a negative pressure, it also eliminates any uncontrolled gas emissions through electrode ports. By 2013, the process was deployed in 35 mills on 3 continents, including the world's largest EAF, a 420-t capacity Tokyo Steel furnace operating at the Tahara plant since June 2010 (Ogawa, Sellan, & Ruscio, 2011). This Danieli design is a twin DC furnace with a diameter of 9.7 m, capacity of 300 t hot metal and 120 t hot heel, maximum power of 175 MW, and tap-to-tap time of 50 min (hence hourly productivity of 360 t), and it consumes 33 m^3/t of oxygen delivered by six jets and a supersonic lance.

Despite these accomplishments, the long-term worldwide trend toward more secondary metal was recently even temporarily reversed because of China's enormous expansion in the smelting of pig iron. The global steel:pig iron ratio was about 0.9 in 1900, it reached 1.41 in 1950, it rose only marginally to 1.48 by the year 2000, and then it fell back to 1.39 by 2010 and increased a bit, to 1.41, by 2013 (WSA, 2015). But the story has been different in the United States thanks to the country's long history of ferrous metallurgy and its large stores of scrap iron.

The US steel:pig iron ratio rose from 0.72 in 1900 to 1.23 in 1950, it stood just above 2.0 in the year 2000, and, despite the fact that the country has become a major scrap exporter, it rose to 2.82 in 2010 and 2.86 in 2013, when nearly three times as much steel came from recycled material than from iron ores. For comparison, steel:pig iron ratios are high in Germany (1.57 in 2013, scrap sourced from domestic stocks) and South Korea (1.61, scrap mostly from imports), still low in China (1.16 in 2013), and, surprisingly, also low in the United Kingdom. About half of the world's 90 steel-producing countries, with most of them being small producers in Europe (Bulgaria, Greece, Portugal), the Middle East, Africa, Latin America, and Asia, now rely only on EAFs and on mostly imported scrap.

CHAPTER 7

Energy Costs and Environmental Impacts of Iron and Steel Production

Fuels, Electricity, Atmospheric Emissions, and Waste Streams

Before 1973, energy supply was just another factor in doing business, because for most industries the total cost of purchased energy was not a particularly onerous burden (at that point, the inflation-adjusted price of oil had been going down for more than two decades), and only the makers of the most energy-intensive products had explicit worries about the cost of fuels and electricity and about the total energy required by their industries. Then the two rounds of oil price rises during the 1970s swiftly elevated energy supply, energy cost, and energy requirements to major economic, social, and political concerns as all industries, and especially all major consumers of fuels and electricity, began to look for ways to reduce their energy consumption and to lower the final energy intensity of their products.

Inevitably, the iron and steel industry has been in the forefront of these efforts. When put into a longer perspective, this has been nothing new: the history of ironmaking can be seen as a continuing quest for higher energy efficiency, and this effort brought typical fuel requirements from almost 200 GJ/t of pig iron in 1800 to less than 100 GJ/t by 1850, to only about 50 GJ/t by 1900 (Heal, 1975), and to less than 20 GJ/t a century later. In relative terms, steel, now requiring less than 20 GJ/t in state-of-the-art mills, is not the most energy-intensive commonly used material: aluminum needs nearly nine times as much energy (175 GJ/t, mostly as electricity), plastics consume mostly between 80 and 120 GJ/t (much of it as hydrocarbon feedstock), copper consumes about 45 GJ/t, and paper's energy cost is up to 30 GJ/t, while lumber and cement need less than 5 GJ/t, and glass goes up to 10 GJ/t.

But, as stressed in this book's first chapter, the world's current annual steel consumption is nearly 20 times that of all four common nonferrous

Still the Iron Age.
DOI: http://dx.doi.org/10.1016/B978-0-12-804233-5.00007-5

metals (Al, Cu, Zn, Pb) combined, and this, combined with the relatively high-energy intensity of steelmaking, makes the industry a leading industrial consumer of fuels and electricity, and hence also a leading emitter of polluting gases and an important contributor to anthropogenic generation of CO_2. Consequently, the first two sections of this chapter will review the recent energy costs of iron- and steelmaking, focusing not only on aggregate needs but also on the requirements of all major specific processes as well as on differences among nations.

Moreover, I will not cite only the latest assessments but will also look at the remarkable evolution of steelmaking's energy intensity: few industries can match ferrous metallurgy in its old, and continuing, quest for reduced energy inputs; to state this in reverse, we would have never achieved current levels of aggregate output and its affordable pricing if we had to spend as much energy per tonne of steel as we did just after WW II (at that time it was about 2.5 times as much as now), and steel output on the order of 1.5 Gt a year would have been a mere fantasy at the 1900 level of more than 50 GJ/t. As one of the classic heavy industries—highly dependent on coke produced from metallurgical coal, consuming large masses of iron ores and fluxing materials, and producing copious solid, liquid, and gaseous wastes—iron and steel was one of the iconic polluters of the late nineteenth century and of the first half of the twentieth century. That reality has changed considerably with the introduction of extensive air pollution controls and with the adoption of new, much more efficient production processes.

Nevertheless, as both a material- and energy-intensive industry still highly dependent on coal, ferrous metallurgy is a major emitter of air pollutants, a leading industrial consumer of water, a minor source of contaminated liquids, a massive producer of solid waste, and a significant source of CO_2. In this chapter's second half, I will assess these impacts and also note the past advances in controlling common pollutants and increasingly common ways to capture and reuse materials produced by the major waste streams, and compare the industry's environmental footprint with that of other leading industrial sectors.

As for the industry's land claims, areas occupied by blast furnaces (BFs) and by buildings housing basic oxygen furnaces (BOFs) and electric arc furnaces (EAFs) are fairly compact, but onsite storage, handling, and processing of iron ore, coal, and fluxing materials claim fairly large areas. Smaller old European and North America iron and steel mills were commonly located inland, albeit preferably on river or lake shores: for example,

Figure 7.1 JFE's Keihin Works south of Tōkyō: coal and iron storage in the foreground; two BFs and their hot stoves on the left. *Reproduced by permission from JFE Steel.*

Andrew Carnegie's Homestead Steel Works on the southern bank of the Monongahela east of Pittsburgh occupied about 112 ha (Carnegie Steel Company, 1912). Modern integrated plants with multiple BFs demand much more space: for example, Shougang Jingtang Iron & Steel occupies a new coastal site of about 2000 ha on the Bohai Bay, and seaside locations on artificial islands reclaimed from bays, the practice pioneered by Japan, have been common for new large enterprises in Asia. The Keihin Works of JFE Steel, just south of Tōkyō, are a typical example of this reclaimed location (Fig. 7.1).

ENERGY ACCOUNTING

Once the era of declining oil prices ended and energy emerged suddenly as the subject of intense interest, it became obvious that we needed to get reliable and comprehensive accounts of energy costs in order to identify the extent and the intensity of energy inputs and to find the best opportunities for the most rewarding savings. In order to undertake such studies, a new discipline of energy analysis emerged during the 1970s (IFIAS, 1974; Thomas, 1979; Verbraeck, 1976). Not surprisingly, new studies concentrated on assessing energy costs of major economic factors, including individual materials (cement, plastics, steel), foodstuffs (corn, wheat), and final products (cars).

Results of these pioneering studies were summarized by Boustead and Hancock (1979), and the assessments generated during the subsequent flourishing of energy analysis during the 1980s can be seen in volumes by Brown, Hamel, and Hedman (1996), Jensen et al. (1998), and Smil, Nachman, and Long (1983). By the mid-1980s, oil prices declined and then remained relatively low and stable for two decades, and this meant that, contrary to early expectations, energy analysis (although it continued to be practiced by some students of energy systems) did not become either an essential tool of energy studies or a major adjunct of economic appraisals. As one of its early pioneers, I have always found it useful, revealing, and highly instructive, but I have been also always aware of its limitations.

Two basic approaches have been used to assess energy costs of products or entire industrial sectors: quantifications based on input–output tables and process analyses. The first option is obviously a variant of commonly used econometric analysis relying on a sectoral matrix of economic activity in order to extract the values of energy inputs and then to convert them into energy equivalents by using representative energy prices. Such a sectoral analysis embraces heterogeneous categories rather than specific products, but it is clearly more suitable for a relatively homogeneous iron and steel industry than for consumer electronics with its huge array of diverse products.

In order to find the energy required to make a specific product (often called embodied energy), it is necessary to perform a process analysis that identifies the sequence of operations required to produce a particular item, traces all important material and direct energy inputs, and finds the values of indirect energy flows attributable to raw materials or finished products entering the process sequence. Process analyses are valuable heuristic and managerial tools, and the gained insights may be used not only to reduce energy requirements but also to rationalize material flows. The choice of system boundaries determines the outcome of process analyses.

In many cases limiting them to direct energy inputs used in the final stage of a specific industrial process may yield satisfactory results. To use a relevant primary ironmaking example, we do not need to account for energy cost of a BF (that is mostly for energy used in smelting the needed steel and producing the refractory materials) in order to account for energy cost of the pig iron it produces. That furnace, with two relinings, could be reducing iron ore for more than half a century, and prorating the energy cost of its construction over the more than 100 Mt of pig iron it will produce during its decades of operation would result in negligibly small additional values that would be also much smaller than the errors associated with even the best accounting for large direct energy inputs. But in other instances, truncation errors arising from the imposition of arbitrary analytical boundaries may be relatively large.

In the case of ironmaking, nontrivial higher order inputs that might be omitted from simple process analyses include the energy costs of mining coal, iron ore, and limestone and the preparation and transportation costs of raw materials. When Lenzen and Dey (2000) looked at energy used by the Australian steel industry, they discovered that lower order needs were just 19 GJ/t, but that the total requirement was 40 GJ/t, which means that truncation error (the omission of higher order energy contributions) doubled the overall specific rate. Similarly, Lenzen and Treloar's (2002) input–output analysis of energy embodied in a four-story Swedish apartment building ended up with a rate twice as large as that established by process analysis by Börjesson and Gustavsson (2000), and the greatest discrepancies concerned structural steel (nearly 17 GJ/t vs. about 6 GJ/t) and plywood (roughly 9 GJ/t vs. 3 GJ/t).

Recent EU rates show a significant difference when excluding a single second-order input: the sequence of BF, BOF and bloom, slab, and billet mill processing is about 55% more energy costly (20.7 GJ/t vs. 13.3 GJ/t) when coke plant energy is included, and the rates rises to more than 25 GJ when the same energy flows are expressed in primary terms, that is, when accounting for fuel energy lost in the generation of fossil-fueled electricity. And my final example of uncertainties inherent in energy analysis concerns different qualities of final products. As I will illustrate, standard energy analyses of modern crude steel show rates around 20 GJ/t, but Johnson et al. (2008) put the total energy cost of austenitic stainless steel (a variety that has been in increasing demand) at the beginning of the twenty first century at 53 GJ/t for the standard process (including small amount of stainless scrap), and at 79 GJ/t for production solely from virgin materials, with nearly half of that total going for extraction and preparation of FeCr, FeNi, and Ni (the steel has 18% Cr and 8% Ni).

These problems of boundary choice and quality disparities are an inherent complication in the preparation of process energy accounts and they are a source of common uncertainties when comparing increasingly common (but still relatively rare) studies of energy costs of leading materials: consequently, there can be no single correct value, but as long as the compared studies use the same, or similar, analytical boundaries and conversions, they offer valuable insights into secular efficiency gains. That is why I will not offer detailed surveys of key studies and their (often misleadingly) precise calculations of energy costs but simply present rounded rates and ranges in order to trace long-term historical trends in using fuels and electricity in the production of iron and steel, both at national and process levels.

A comprehensive energy analysis requires tracing at least direct energy inputs, including all fuels and electricity, and preferably both direct and indirect energy requirements, particularly for those processes whose material inputs require considerable energy investment and where electricity is a large or dominant form of purchased (or in-plant generated) energy. While comprehensive accounting is necessary to produce realistic estimates of total energy costs, close attention must be paid to dominant inputs where accounting errors may be easily larger than totals supplied by minor form of energy used in a specific process: in ironmaking this means, obviously, coming up with accurate assessments of energy costs of coke production and other fuels used in BFs.

In mass terms, these fuels (dominated by coal-derived coke and also including coal dust, natural gas, and fuel oil) are the second largest input in the production of pig iron: as already noted, typical requirements for producing a tonne of the metal in a BF are 1400 kg of iron ore, 800 kg of coal (indirectly for coking, directly for injection), 300 kg of limestone, and 120 kg of recycled metal (WSA, 2012b). Hydrocarbons have a distinctly secondary position, but direct reduction of iron using inexpensive natural gas should be gaining in importance. Electricity (be it fossil-fuel generated, nuclear, or hydro) is a comparatively minor energy input in iron ore reduction in BFs, but it is indispensable for energizing EAF-based steelmaking and for operating continuous casting and rolling processes. And given the volumes of hot gases and water generated by ironmaking and steelmaking, it is also important to account for energy values of waste streams available for heat recovery.

In aggregate monetary terms, energy use in steelmaking ranges between 20% and 40% of the final cost of steel production; for example, when using long-term prices Nucor puts the cost of energy for operating a BF at 22% of the pig iron costs (Nucor, 2014), while a Japanese integrated steelmaker (with its own coking and sintering plants using imported coal and iron ore) spends 35% of its total (and about 38% of its variable) cost on energy. Obviously, these relatively high-energy costs would have been a rewarding target for reduction even if the industry would not have been affected by rising prices of coal, crude oil, natural gas, and electricity—and the post-1973 increases (as well as unpredictable fluctuations) in energy cost had only strengthened the quest for lower energy intensity of iron and steel production, resulting in some impressive fuel and electricity savings.

In surveying these gains, one should always specify the national origins (there are appreciable differences among leading steel-producing countries), make it clear which energy rate is calculated, quoted, or estimated, and to what year they apply, and note if the cited rates are national averages, typical performances in the industry, or the best performances of the most modern operations, and if they refer to the entire steelmaking process or only to its specific parts; unfortunately, all too often these are explained only partially, or they are entirely assumed, leaving a reader with rates that may not be comparable.

The most common difference is between the accounts that use only direct energy and those expressing the costs in terms of primary energy (including energy losses in generating electricity and converting fuels). This will make the greatest difference in the case of processes heavily dependent on electricity: in Europe, recent direct energy use by an EAF is 2.5 GJ/t of steel, primary energy of that input is about 6.2 GJ/t, and the two rates for energy used by a hot strip mill are, respectively, 1.7 and 2.4 GJ/t (Pardo, Moya, & Vatopoulos, 2012). In the case of energy use by BFs, the most common accounting difference arises from imposing analytical boundaries: some analyses include the energy cost of cokemaking, but most of them omit it.

ENERGY COST OF STEELMAKING

Because iron and steel industry has been always a rather energy-intensive enterprise with continuing interest in managing and reducing energy inputs, we have fairly accurate accounts, including detailed retrospective appraisals, that allow us to trace the sector's energy consumption trends for the entire twentieth century and, in a particularly rich detail, for the past few decades (Dartnell, 1978; De Beer, Worrell, & Blok, 1998; Hasanbeigi et al., 2014; Heal, 1975; Leckie, Millar, & Medley, 1982; Smithson &

Sheridan, 1975; Worrell et al., 2010). I will start with energy costs of pig iron smelting in BFs, and then proceed to electricity expenditures for BOFs, EAFs, and rolling before summing up the process totals. But before reviewing these rates, I will first introduce the minimum energy requirements of common steelmaking processes, summarized by Fruehan et al. (2000), and compare them with the best existing practices. Contrasting these two rates makes it possible to appreciate how closely they have been approached by the combination of continuing technical advances aimed at maximizing energy efficiency of key steelmaking processes.

Inherently high-energy requirements for reducing iron oxides and producing liquid iron in BFs dominate the overall energy needs in integrated steelmaking. In the US steel industry, with its high share of secondary steelmaking, about 40% of all energy goes into ironmaking (including sintering and cokemaking), nearly 20% into BOF and EAF steelmaking, and the remainder into casting, rolling, reheating, and other operations (AISI, 2014). In India, where primary metal smelting dominates, about 70% of the sector's energy goes for ironmaking (BF 45%, coking 15%, and sintering 9%), 9% for steelmaking, 12% for rolling, and 10% for other tasks (Samajdar, 2012).

Iron ore (Fe_2O_3) reduction requires at least 8.6 GJ/t, and the absolute minimum of producing pig iron in BF (5% C, tap temperature 1450 °C) is 9.8 GJ/t of hot metal; a more realistic case must include the energy needed for the formation of slag and for a partial reduction of SiO_2 and MnO (hot metal containing 0.5% Si and 0.5% Mn), as well as the effect ash in metallurgical coke: slag effect increases the minimum requirements to 10.27 GJ/t, and slag and coke ash effect result in a slightly higher rate of 10.42 GJ/t. In contrast, Worrell et al. (2008) put the best commercial performance for BF operation at 12.2 GJ (12.4 GJ in primary energy terms), and Worrell et al. (2010) offer the range of 11.5–12.1 GJ/t.

As for the inputs, the absolute theoretical minimum for ore agglomeration is 1.2 GJ/t of output, that is, 1.6 GJ/t of steel, while Fruehan et al. (2000) put actual demand at 1.5–1.7 GJ/t of output and 2.1–2.4 GJ/t of steel. Worrell et al. (2008) estimated the best actual rate at 1.9 GJ/t (2.2 GJ/t in terms of primary energy), Worrell et al. (2010) quoted the range of 1.62–1.85 GJ/t, and according to Outotec (2015b), the world leader in iron ore beneficiation, the process needs 350 MJ of heat per tonne of pellets for magnetite ores and 1.5 GJ/t for limonites, and, depending on the ore and plant capacity, an additional 25–35 kWh per tonne for mixing, balling, and induration, for totals between 0.6 and 1.9 GJ/t of pellets.

Coke output in modern plants amounts to about 0.77 t per tonne of coal input; the remainder consists of captured volatiles used either as fuel or chemical feedstocks. Captured coke gas has a relatively high-energy density gas as it contains 11.8–14.5 GJ/t of coke (or 4–5 GJ/t of produced steel). After taking this valuable energy output into account, the minimum net energy required for cokemaking is about 2 GJ/t or 0.8 GJ/t of steel (Fruehan et al., 2000), while actual recent performances range between 5.4 and 6.2 GJ/t of coke, that is, 2.2–4.6 GJ/t of steel.

Reconstructions of overall past energy requirements show that at the beginning of the twentieth century direct energy needed for BF smelting (all but a tiny share of it as metallurgical coke, but excluding the energy cost of coking) was between 55 and 60 GJ/t of pig iron, and by 1950 that range was reduced to 35–45 GJ/t. By the early 1970s, common Western performance of BF ironmaking was about 30 GJ/t, and the best rates were no better than 25 GJ/t, but then the OPEC-engineered oil price rise of 1973–1974 and its second round in 1979–1980 led to an accelerated progress in energy savings. By the end of the twentieth century, the net specific energy requirement of state-of-the-art BFs was no more than 15 and as little as 13 GJ/t. That was as much as 50% less than in 1975 and, even more remarkably, it was as little as 25% above the minimum energy inputs needed to produce pig iron from coke-fueled smelting of iron ore, while common performances were still 40–45% above the energetic minimum.

As already explained in the previous chapter, these impressive gains in the production of pig iron were due to the combination of many technical fixes, and the principal savings attributable to specific improvements are as follows (IETD, 2015; USEPA, 2012). Dry quenching of coke may save more than 0.25 GJ/t, recovery of sintering heat saves 0.5 GJ/t, and the capture and combustion of top gases may reduce total energy use by up to 0.9 GJ/t of hot metal. Increased coal injection saves about 3.75 GJ/t of injected fuel; every tonne of injected coal displaces 0.85–0.95 t of coke, and the fuel savings are nearly 0.8 GJ/t of hot metal. Increased hot blast temperatures save up to 0.5 GJ/t, and heat recuperation from hot blast stoves cuts demand by up to 0.3 GJ/t. Higher BF top pressures reduce coke rates and allow more efficient electricity generation by recovery turbines, yielding as much as 60 kWh/t of hot metal. And improved controls of the hot stove process may save up to 0.04 GJ/t.

Steelmaking does not present such large opportunities for energy savings in absolute terms, but relative reductions of fuel and electricity requirements have been no less impressive than in ironmaking, with much of the reduced energy intensity due to the displacement of OHFs by BOFs in integrated enterprises and by EAFs in mini-mills. Steelmaking in BOF, using hot pig iron and scrap, involves a highly exothermic oxygenation of carbon, silicon, and other elements, and hence the process is a net source of energy even after taking into account the about 600 MJ (as electricity) needed to make oxygen used in producing a tonne of hot metal. Compared to OHFs (they needed about 4 GJ/t) the overall saving will thus be more than 3 GJ/t, and the final energy cost of BOF steel will thus be essentially the cost of the charged hot pig iron. Depending on the amount of scrap melted per tonne of hot metal (typically between 30 and 40 kg) and on its specific composition (assuming 5% C and 0.5% Si, presence of coke ash, and 20–30% FeO in the slag), the energy cost of crude BOF steel would be no less than 7.85 and up to 8.21 GJ/t (Fruehan et al., 2000).

Theoretical minima to produce steel by melting scrap in EAFs vary only slightly with the composition of the charge metal and the share of FeO in slag, between 1.29 and 1.32 GJ/t, but because large volumes of air (up to 100 m^3/t) can enter the furnace (mainly through its door), the heating of entrained N_2 would raise the total demand to 1.58 GJ/t (Fruehan et al., 2000). In contrast, the recently cited averages for large-scale production have ranged from about 375 to 565 kWh/t (Ghenda, 2014). Capacity of 100 t/heat and tap-to-tap time of 40 min would translate to between 3.8 and 5.8 GJ/t in terms of primary energy. Worrell et al. (2010) use the US mean of 4.5 GJ/t and that is also the approximate average cited by Emi (2015), compared to less energy-intensive melting of scrap in BOF that needs only about 3.9 GJ/t.

The electricity demand of large EAF furnaces presents a challenge for the reliability of supply and the stability of grids, even with the most efficient designs. SIMETAL's Ultimate EAF requires only 340 kWh/t of steel (its melting power is 125–130 MW), which means that in 1 day (with 48 heats of 120 t) it needs 1.95 GWh of electricity or—using the average annual household electricity consumption of 10.9 MWh (USEIA, 2015)—as much as a city with 65,000 households (i.e., with roughly 165,000 people). Additional investment may be needed to prevent delivery problems and to assure the reliability of supply for other consumers in the area with a number of these extraordinarily electricity-intensive devices. As already noted, the two effective steps toward reducing EAF energy requirements are charging of hot pig iron and preheating of scrap.

The world's best practices in casting and rolling are as follows: continuous casting and hot rolling, 1.9 (2.5) GJ/t, and cold rolling and finishing, 1.5 (2.3) GJ/t (Worrell et al., 2008). Replacing traditional rolling of semi-finished products from ingots (requiring 1.2 GJ/t) by continuous casting (whose intensity is just 300 MJ/t) saves almost 1 GJ/t. There is, obviously, a substantial difference between energy requirements for cold rolling and hot rolling that needs reheating of cast metal. For flat carbon steel slabs, the difference is 50-fold (17 GJ/t vs. 850 GJ/t), for stainless steel slabs it is about 17-fold, about 50 versus nearly 900 GJ/t (Fruehan et al., 2000). The world's best practices now require (in primary energy terms) 2.2 GJ/t for hot-rolling strip steel, 2.4 GJ/t for hot-rolling bars, and 2.9 GJ/t for hot-rolling wires (Worrell et al., 2008). Thin slab casting requires about 1 GJ/t, but strip casting consumes only 100–400 MJ/t.

Making specialty steel is more energy intensive. Production of the most common variety of stainless steel (18-8, with 18% Cr and 8% Ni) using EAF (charged with 350 kg of steel and 400 kg of stainless scrap) and argon oxygen decarburization (AOD) sequence requires at least 1.21 GJ/t. All of these values refer only to direct input of electricity and exclude losses in generating and transmitting electricity, as well as all second-order inputs including the energy cost of the furnace itself and its replacement electrodes and refractories. Actual electricity (direct energy) use in modern EAFs is about 2.5 GJ/t (even with high average 40% conversion efficiency that means 6.25 GJ/t of primary energy in all instances where electricity is generated by the combustion of fossil fuels in central power stations).

Energy savings resulting from the adoption of new processes and from gradual improvements of old practices have eventually added up to impressive reductions per unit of final product. The total energy requirement for the UK's finished steel was cut from about 90 GJ/t in 1920 to below 50 GJ/t by 1950, during the decades relying on BF, OHF, and traditional casting. By 1970, the best integrated mills still using OHF needed 30–45 GJ/t of hot metal, but by the late 1970s (with higher shares of BOF and CC), nationwide means in both the United Kingdom and the United States were less than 25 GJ, and the combined effects of advances in integrated (BF–BOF–CC) steelmaking (with higher reliance on EAF) reduced typical energy cost to less than 20 GJ/t by the early 1990s, with more than two-fifths of savings due to pig iron smelting, a few percent claimed in BOFs, and the remainder in rolling and shaping (De Beer, Worrell, & Blok, 1998; Leckie et al., 1982).

In the United States, the final energy use per tonne of crude metal shipped by the steel industry declined from about 68 G/t in 1950 to just over 60 GJ/t in 1970 and to 45 GJ/t in 1980, and then, with the shift toward mini-mills and EAF, it fell by nearly three-quarters in three decades: by the year 2000, the US nationwide rate was 17.4 GJ/t (USEPA, 2012). A detailed study of the sector's energy intensity (including all cokemaking, agglomeration, ironmaking, steelmaking, casting, hot and cold rolling, and galvanizing and coating) put the nationwide mean at 14.9 GJ/t in 2006 (Hasanbeigi et al., 2011); and by 2010 the rate was just 11.8 GJ/t, with the industry reducing the average energy need by nearly 75% in three decades. In 2005, the American Iron and Steel Institute published a roadmap for transformation of steelmaking processes: SOBOT (saving one barrel of oil per ton) should lower the overall energy cost from an equivalent of 2.07 barrels of oil per ton in 2003 to just 1.2 barrels a ton in 2025 (AISI, 2005). The comparison assumes 49% EAF share in 2003 and 55% EAF share in 2025, and the 2025 rate would be equivalent to about 9.7 GJ/t.

For China, the world's largest steel producer, we have several recent studies. Guo and Xu (2010) put the national average of energy requirements for steelmaking at 22 GJ/t in the year 2000 and 20.7 GJ/t in 2005, with 2004 rates for coking at 4.1, for ironmaking at 13.5, for EAFs at 6.0, and for rolling at 2.6 GJ/t. Chen, Yin, and Ma (2014) found that China's average energy requirement in key iron and steel enterprises (hence not a true national average) declined by nearly 20% between 2005 and 2012 when it was 17.5 GJ/t, and that there were substantial differences between the average and the most and the least efficient enterprises: in 2012 the relevant rates were 11.6, 9.9, and 13.5 GJ/t for ironmaking, and 2, 0.7, and 5.3 GJ/t for steelmaking in EAFs.

Analyses of energy use by Canada's iron and steel industry show a less impressive decline, from the mean of 20.9 GJ/t of crude steel in 1990 to 17.23 GJ/t in 2012, a reduction of about 20% in 22 years (Nyboer & Bennett, 2014). And reductions in specific energy consumption in German steelmaking have been even smaller, amounting to just 6.3% between 1991 and 2007, with about 75% of those gains explained by a structural shift away from BF/BOF to a higher share of EAF production (Arens, Worrell, & Schleich, 2012). Gains in BF efficiency have been only 4%, with the heat rate declining from 12.5 to 12 GJ/t in 16 years. Average energy consumption of German iron and steel industry in 2013 was 19.23 GJ/t when measured in terms of finished steel products (21% reduction since 1990) and 17.42 GJ/t in terms of crude steel (Stahlinstitut

VDEh, 2014). And JFE Steel, Japan's second largest steel producer, lowered its specific energy use from 28.3 GJ/t steel in 1990 to 23.3 GJ/t in 2006, and the same rate applied in 2011 (Ogura et al., 2014).

There used to be substantial intranational (regional) differences between energy requirements of steelmaking in large economies, but the diffusion of modern procedures has narrowed the gaps. At the same time, differences in nationwide average of energy costs in steelmaking will persist. Higher rates are caused by less exacting operation and maintenance procedures as well by the low quality of inputs, such as India's inferior coking coals (with, even after blending, 12–17% of ash compared to 8–9% elsewhere), or iron ores requiring energy-intensive beneficiation. As a result, in comparison to practices prevailing among the world's most efficient producers, India's cokemaking consumes 30–35% more energy; iron ore extraction and preparation has energy intensity 7–10% higher: Samajdar (2012) puts the aggregate average range at 27–35 GJ/t.

China's steelmaking used to be very inefficient: during the early 1990s the mean energy cost was 46–47 GJ/t of metal, and after rapid additions of new, modern capacities the rate fell to a still high 30 GJ/t by the year 2000 (Zhang & Wang, 2009). Continuing improvements and the unprecedented acquisition of large, modern, efficient plants during the past two decades resulted in further energy intensity reduction, but a detailed comparison of energy costs of steel in the United States and China showed that by 2006 the nationwide mean for China's crude steel production (23.11 GJ/t) was still 55% above the US average of 14.9 GJ/t (Hasanbeigi et al., 2014).

But national means of energy costs reflect not only many specific technical accomplishments (or their lack of) but also the shares of major steelmaking routes: countries with higher shares of scrap recycling have significantly lower national means. When Hasanbeigi et al. (2014) performed another analysis that assumed the US share of EAF production to be as low as in China (just 10.5% in 2006, obviously limited by steel scrap availability in a country whose metal stock began to grow rapidly only during the 1990s), the US mean rose to 22.96 GJ/t, virtually identical to the Chinese mean (and a hardly surprising finding given the fact that most of China's steelmaking capacity was, as just noted, installed after the mid-1990s).

Differences arising from the choice of analytical boundaries and conversion factors are well illustrated by an international comparison of steel's energy cost published by Oda et al. (2012). In their macrostatistical approach, they excluded energy cost of ore and coal extraction and their transportation to steel mills, included the cost of cokemaking and ore agglomeration and all direct and indirect energy inputs into blast, oxygen, and electric furnaces, casting, and rolling, and converted all electricity at a rate of 1 MWh = 10.8 GJ. Their results are substantially higher than for all other cited estimates: their average for the BF–BOF route in the United States, 35.5 GJ/t, is three times the US rate calculated by Hasanbeigi et al. (2011). Other rates are 28.8 GJ/t for the EU, 25.7 GJ/t for Japan, 30.5 GJ/t for China, and 30 GJ/t for India (both rates about 15% lower than in the United States!), but 65 GJ/t for Russia and a worldwide mean of 32.7 GJ/t, all for the year 2005.

Finally, a few key comparisons of the industry's energy requirements. My approximate calculation is that in 2013 worldwide production of iron and steel claimed at least 35 EJ of fuels and electricity, or less than 7% of the total of the world's primary energy supply; for comparison, Laplace Conseil (2013) put the share at about 5% for 2012, compared to 23% for all other industries, 27% for transportation, and 36% for residential use and services. In either case that makes iron and steel the world's largest energy-consuming industrial sector, further underscoring the need for continuing efficiency gains. In terms of specific fuels, the sector's energy use claims 11% of all coal output, only about 2% of all natural gas, and 1% of electricity (use of liquid hydrocarbons is negligible).

At the same time, it is necessary to appreciate the magnitude of the past improvements. If the sector's energy intensity had remained at its 1900 level, then today's ferrous metallurgy would be claiming no less than 25% of all the world's primary commercial energy. And if the industry's performance had remained arrested at the 1960s level (when it needed 2.5 times as much energy as it does now), then the making of iron and steel would require at least 16% of the world's primary energy supply.

National shares depart significantly from the global mean, reflecting both the magnitudes of annual output and the importance of other energy-consuming sectors. In 1990, Japan's iron and steel industry consumed 13.6% of the nation's primary energy; the share was down to 10.7% in the year 2000 and a marginally better 10.3% in 2010, indicating a still relatively high importance of ferrous metallurgy in the country's economy (JISF, 2015). Energy consumption in the US iron and steel industry peaked in 1974 at about 3.8 EJ or roughly 5% of the country's total primary energy use. Post-1980 decline of pig iron smelting, the country's high rates of energy use in households, transportation, and services, and improvements in industrial energy intensity combined to lower the ferrous metallurgy's overall energy claim to only 1.3% of all primary energy by 2013.

Similarly, in Canada the share of iron and steel industry in national primary energy use declined from 2.5% in 1990 to 1.6% in 2010 (Nyboer & Bennett, 2014). In contrast, China's primary energy demand is still dominated by industrial enterprises whose output has made the country the world's largest exporter of manufactured goods and has provided inputs for domestic economy that, until 2013, grew at double-digit rates. But because of unprecedented post-1995 expansion of China's steelmaking, its energy claim has translated into an unusually high share of overall energy use: it rose from just over 10% in 1990 to nearly 13% by the year 2000 (Zhang & Wang, 2009), Guo and Xu (2010) put it at 15.2% for the year 2005, and in 2013 it was, according to my calculations, nearly 16%, much higher than in any other economy.

Given the substantial gains achieved during the past two generations (recall how closely some of the best practices have now approached the theoretical minima), future opportunities for energy savings in iron and steel industry are relatively modest, but important in aggregate. Details of these opportunities are reviewed and assessed, among many others, by AISI (2005), Brunke and Blesl (2014), Ogura et al. (2014), USEPA (2007 and 2012), and Worrell et al. (2010). Their deployment is still rewarding even in Japan, the country with the highest overall steelmaking efficiency (Tezuka, 2014). Besides such commonly used energy-saving measures as dry coke quenching, recovery of heat in sintering or BF top pressure gas turbines, Japanese steelmakers have also introduced a new scrap-melting shaft furnace (20 m tall, 3.4 m diameter, 0.5 Mt/year annual capacity) and a new sintering process where coke breeze is partially replaced by natural gas (Ogura et al., 2014).

AIR AND WATER POLLUTION AND SOLID WASTES

Ferrous metallurgy offers one of the best examples of how a traditional iconic polluter, particularly as far as the atmospheric emissions were concerned, can clean up its act, and do so to such an extent that it ceases to rank among today's most egregious offenders. But environmental impacts of iron- and steelmaking go far beyond the release of airborne pollutants, and I will also review the most worrisome consequences in terms of waste disposal, demand for water, and water pollution. And while iron and steel mills are relatively compact industrial enterprises that do not claim unusually large areas of flat land (many of them, particularly in Japan, are located on reclaimed land), extraction of iron ores has major local and regional land use impacts in areas with large-scale extraction, above all in Western Australia and in Pará and Minas Gerais in Brazil.

All early cokemaking, iron smelting, and steelmaking operations could be easily detected from afar due to their often voluminous releases of air pollutants whose emissions were emblematic of the industrial era: particulate matter (both relatively coarse with diameter of at least 10 μm, as well as fine particles with diameter of less than 2.5 μm that can easily penetrate into lungs), sulfur dioxide (SO_2), nitrogen oxides (NO_x, including NO and NO_2), carbon monoxide (CO) from incomplete combustion, and volatile organic compounds. Where these uncontrolled emissions were confined by valley locations with reduced natural ventilation, the result was a chronically excessive local and regional air pollution: Pittsburgh and its surrounding areas were perhaps the best American illustration of this phenomenon.

Recent Chinese rates and totals illustrate both the significant contribution of the sector to national pollution flows and the opportunities for effective controls. Guo and Xu (2010) estimated that the sector accounted for about 15% of total atmospheric emissions, 14% of all wastewater and waste gas, and 6% of solid waste, and they put the nationwide emission averages in the year 2000 (all per tonne of steel) at 5.56 kg SO_2, 5.1 kg of dust, 1.7 kg of smoke, and 1 kg of chemical oxygen demand (COD). But just 5 years later spreading air and water pollution controls and higher conversion efficiencies reduced the emissions of SO_2 by 44%, those of smoke and COD by 58%, and those of dust by 70%.

Particulates are released at many stages of integrated steelmaking, during ore sintering, in all phases of integrated steelmaking as well as from EAFs and from DRI processes, but efficient controls (filters, scrubbers, baghouses, electrostatic precipitators, cyclones) can reduce these releases to small fractions of the uncontrolled rates (USEPA, 2008). Sintering of ores emits up to about 5 kg/t of finished sinter, but after appropriate abatement maximum EU values in sinter strand waste gas are only about 750 g of dust per tonne of sinter, and minima are only around 100 g/t, but there are also small quantities of heavy metals, with maxima less than 1 g/t of sinter and minima of less than 1 mg/t (Remus et al., 2013). In the United States, modern agglomeration processes (sintering and pelletizing) emit just 125 and up to 250 g of particulates per tonne of enriched ore (USEPA, 2008). Similarly, air pollution controls in modern coking batteries limit the dust releases to less than 300 g/t of coke and SO_x emissions (after desulfurization) to less than 900 g/t, and even to less than 100 g/t.

Smelting in BFs releases up to 18 kg of top gas dust per tonne of pig iron, but the gas is recovered and treated. Smelting in BOFs and EAFs can generate up to 15–20 kg of dust per tonne of liquid steel, but modern controls keep the actual emissions from BOFs to less than 150 g/t or even to less than 15 g/t, and from EAFs to less than 300 g/t (Remus et al., 2013). Long-term Swedish data show average specific dust emissions from the country's steel plants falling from nearly 3 kg/t of crude steel in 1975 to 1 kg/t by 1985 and to only about 200 g/t by 2005 (Jernkontoret, 2014).

But there is another class of air pollutants that is worrisome not because of its overall emitted mass but because of its toxicity. Hazardous air pollutants originate in coke ovens, BFs, and EAFs. Hot coke gas is cooled to separate liquid condensate (to be processed into commercial by-products, including tar, ammonia, naphthalene, and light oil) and gas (containing nearly 30% H_2 and 13% CH_4) to be used or sold as fuel. Coking is a source of particulates, volatile organic compounds, and polynuclear aromatic hydrocarbons: uncontrolled emissions per tonne of coke are up to 7 kg of particulate matter, up to 6 kg of sulfur oxides, around 1 kg of nitrogen oxides, and 3 kg of volatile organics. Ammonia is the largest toxic pollutant emitted from cokemaking, and relatively large volumes of hydrochloric acid (HCl) originate in pickling of steel, when the acid is used to remove oxide and scale from the surface of finished metal. Manganese, essential in ferrous metallurgy due to its ability to fix sulfur, deoxidize, and help in alloying, has the highest toxicity among the released metallic particulates, with chromium, nickel, and zinc being much less worrisome.

But, again, modern controls can make a substantial difference: USEPA's evaluations show that the sector's toxicity score (normalized by annual production of iron and steel) declined by almost half between 1996 and 2005 and that the mass of all toxic chemicals was reduced by 66% (USEPA, 2008). And these improvements have continued since that time. Water used in coke production and for cooling furnaces is largely recycled, and wastewater volumes that have to be treated are relatively small, typically just 0.1–0.5 m^3/t of coke and 0.3–6 m^3/t of BOF steel. Wastewater from BOF gas treatment is processed by electrical flocculation while mill scale and oil and grease have to be removed from wastewater from continuous casting. EAFs produce only small amounts of dusts and sludges, usually less than 13 kg/t of steel (WSA, 2014a). Dust and sludge removed from escaping gases have high iron content and can be reused by the plant, while zinc oxides captured during EAF operation can be resold.

But solid waste mass generated by iron smelting in BFs is an order of magnitude larger, typically about 275 kg/t of steel (extremes of 250–345 kg/t), and steelmaking in BOFs adds another 125 kg/t (85–165 kg/t). The BF/BOF route thus leaves behind about 400 kg of slag per tonne of metal, and the global steelmaking now generates about 450 Mt of slag a year—and yet this large mass poses hardly any disposal problems. Concentrated and predictably constant production of the material and its physical and chemical qualities, that make it suitable for industrial and agricultural uses, mean that slag is not just another bothersome waste stream but a commercially useful by-product.

The material is marketed in several different forms which find specific uses (NSA, 2015; WSA, 2014b). Granulated slag is produced by rapid water cooling; it is a sand-like material whose principal use is incorporation into standard (Portland) cement. Air-cooled slag is hard, dense, and chunky material that is crushed and screened to produce desirable sizes used as aggregates in precast and ready-mixed concrete, in asphalt mixtures or as a railroad ballast and permeable fill for road bases, in septic fields, and for pipe beds. Pelletized (expanded) slag resembles a volcanic rock, and its lightness and (when ground) excellent cementitious properties make it a perfect aggregate to make cement or to be added to masonry. Expanded slag is now widely used in the construction industry, and Lei (2011) reported that in 2010 China's cement industry used all available metallurgical slag (about 223 Mt in that year). Brazilian figures for 2011 show 60% of slag used in cement production, 16% put into road bases, and 13% used for land leveling (CNI, 2012).

High content of free lime prevents the use of some slag in construction, but after separation both materials become usable, with lime best used as fertilizer. Because of its high content of basic compounds (typically about 38% CaO and 12% MgO), ordinary slag is an excellent fertilizer used to control soil pH in field cropping as well as in nurseries and parks and for lawn maintenance and land recultivation; slag also contains several important plant micronutrients, including copper, zinc, boron, and molybdenum.

LIFE CYCLE ASSESSMENTS

Life cycle assessment (LCA) is the most comprehensive approach to the compilation and evaluation of potential environmental impacts of entire product systems throughout their, often complex, history (ISO, 2006). LCA is particularly revealing as it allows us to compare environmental impacts in their entirety rather than, misleadingly, choosing a single (albeit the most important) variable or focusing only on a segment of a complex production process. Consequently, an LCA of steel should start with raw material extraction, include all relevant ironmaking and steelmaking processes, follow the material flow through manufacturing and use, and look at the recycling and disposal of obsolete products (WSA, 2011c).

Assessed variables range from measures of toxicity to humans and ecotoxicity of water and sediments to nutrient loading (eutrophication), acidification, photochemical ozone creation potential (POCP), and global warming potential. Again, as is the case with energy analyses, data limitations and differences in analytical boundaries and conversion ratios may complicate the comparability of specific results. And, obviously, when comparing specific studies it is necessary to look at identical, or at least very similar, product categories, for example, at heavy-duty structural steel, or at least at a broad category of structural steel.

There are now many published LCA values for steel. Besides LCAs for specific steel products or applications—for example, for truck wheels (PE International, 2012), tubular wind towers (Gervásio et al., 2014), or bridges (Hammervold, Reenaas, & Brattebø, 2013)—there are also assessments for average environmental impacts of national steel production—such as Thakkar et al. (2008) for India and Burchart-Korol (2013) for Poland—and national LCAs offering specific values for a wide range of finished products. A Canadian LCA (Markus Engineering Services, 2002) provided highly disaggregated impact values for nails; welded wire mesh and ladder wire; screws, nuts, and bolts; heavy trusses; open web joists; rebar rods; HSS, tubing; hot rolled sheet; cold rolled sheet; galvanized sheet; galvanized deck; and galvanized studs. And there are also LCAs for alternative resources in ironmaking (Vadenbo, Boesch, & Hellweg, 2013).

Not surprisingly, given the commonalities (or outright identities) of major production processes, these published rates, while displaying national and regional differences, are generally in fairly close agreement, but care must be taken to compare the values for the same production routes (not for a product made by the BF/BOF route and another one using EAF) and for similar time periods. National averages and international appraisals of typical impact values suffice for the first-order comparisons with competing materials. LCAs of steel in Western economies show that advancing air and water pollution controls have removed the industry from the list of the most worrisome emitters, and they also indicate generally low or very low impacts in terms of human toxicities and ecotoxicities (on the order of 0.05 mg/t of crude steel), acidification (2–4 kg of SO_2 equivalent per tonne of steel), eutrophication, and POCP.

Carbon emissions: But, not surprisingly, LCAs of steel production also confirm that the sector's high reliance on coal has made it an important emitter of greenhouse gases. The industry emits mostly CO_2, and only small volumes of CH_4 are released during coking—typically a mere 0.1 g/t of coke (IPCC, 2006)—sintering, and BF operation. Generation of CO_2 is, of course, at the core of iron oxide reduction in BFs, as the oxides of iron react with CO produced by the combustion of coke and coal and produce pig iron and CO_2. In addition, the calcination of carbonate fluxes produces CaO, MgO, and CO_2. CO_2 emissions during steelmaking are comparatively modest because pig iron contains no more than about 4% of carbon to be oxidized, while finished steel retains some of it.

These iron- and steelmaking CO_2 emissions cannot be eliminated as long as we rely on BF and BOF, and the only way to control them would be their capture and permanent storage. In contrast, CO_2 emissions associated with ore mining, agglomeration, coking, and electricity consumption can be reduced by improving efficiencies of relevant conversions. And, of course, higher rates of scrap-based steelmaking are another source of reducing CO_2 emissions. Specific emissions, all cited in tCO_2/t of liquid steel, are: from 1.4 to 2.2 t for integrated steel mills in the West (typically about 1.8–2.0), but as much as 3.5 in India; 1.4 to 2.0 t for natural gas-based direct reduction processes, but as much as 3.3 in India for DRI using low-quality coal; and just 0.4 to 1.1 t for scrap-based smelting in EAFs (Gale & Freund, 2001; IEA, 2012; OECD, 2001; USEPA, 2008).

Chen, Yin, and Ma (2014) put the 2012 mean at 2.3 tCO_2/t of metal for the BF/BOF route and 1.7 tCO_2/t of metal for EAFs (this high rate is due to China's overwhelmingly coal-based electricity generation). But Thakkar et al. (2008) put direct emissions at only 2.01 tCO_2/t for some of India's large integrated steel mills, while according to Burchart-Korol (2013) the average Polish BF/BOF route emissions are as high as 2.46 t and EAF emissions are at 913 kg of CO_2/t. Typical direct European emissions listed by Pardo et al. (2012) are (all in tCO_2/t of crude steel) 2.27 for BFs, about 0.2 for BOFs, 0.24 for EAFs, between 0.08 and 0.09 for different hot mills, and just 0.008 (8 kg) for cold mills. CO_2 emissions in Germany in 2013 averaged 1.466 t/t of product when measured in terms of finished steel products (22% reduction since 1990) and 1.328 t/t in terms of crude steel (Stahlinstitut VDEh, 2014). Average specific CO_2 emissions of Canada's iron and steel industry show a decline from 2.13 t/t of output in 1990 to 1.72 t in 2011 (Nyboer & Bennett, 2013).

About 70% of all emissions from the BF/BOF sequence originate in preparing the charges into BFs and in their now prolonged operation. All of the following rates are expressed in kg CO_2 per tonne of steel, and the shares of CO_3 in the overall volumes of generated gases are in parentheses (IEA, 2012). Preparation of self-fluxing sinters emits mostly between 200 and 350 kg CO_2 (290 kg might be a good average, with CO_2 just 5–10% of the gas volume); lime kilns preparing CaO flux release 57 kg (30%); modern coking keeps the emissions below 300 kg (average 285 kg, 25%); generating hot blast in stoves adds about 330 kg (25%); and BF gas carrying away the products of iron ore reduction amounts to 1255 kg of CO_2 equivalent, and their combustion in an adjacent electricity-generating plant releases about 700 kg/t (CO_2 being about 20% of the flue gas). Finally, releases attributable to hot rolling and to BOFs add, respectively, about 85 and 65 kg. The total CO_2 emissions thus come to at least 1.8 t per tonne of rolled coil (to be used in making cars or appliances).

Calculating the global total of the steel industry's CO_2 emissions and expressing it as a share of global anthropogenic releases of the gas are exercises in unavoidable approximations. For example, IPCC (2007) put the industry's share at 6–7% of anthropogenic CO_2 emissions, and IEA (2008) put it at 4–5%. Assuming global averages of 2.1 t CO_2 for integrated steelmaking (dominated by Chinese production) and 1 t CO_2 for EAFs would yield 2012 emissions (with roughly 1.1 Gt of integrated and 0.45 Gt of EAF steel output) of 2.75 Gt. This would have been nearly 8% of total anthropogenic CO_2 emissions in that year (about 35.6 Gt), more than 8% of all emissions attributable to the combustion of fossil fuels (about 33 Gt), and about 25% of all emissions from industries (11.5 Gt).

My simple calculations are confirmed by the Steel CO_2 Model by McKinsey (2014): it attributes 8% of the world's 2011 CO_2 emissions to steel (direct contribution of 5.6%, electricity generation 0.7%, and mining of ores, coal, and limestone 1.7%). That works out to about 31% of all industrial emissions estimated by McKinsey. Similarly, Hidalgo et al. (2003) put the share of the sector's CO_2 emissions at about 28% of the EU's total industrial releases. The iron and steel industry thus contributes twice as much as the emissions from chemical syntheses, and about 60% more than the production of cement and 45% more than the world's oil and gas industry (electricity generation, with nearly 25%, is the largest contribution resulting from the combustion of fossil fuels).

There are several effective ways to achieve considerable reductions of specific CO_2 emissions, mainly thanks to the combination of the just reviewed decline in energy intensity of pig iron production and capture and reuse of CO_2-rich BF gases, and in many countries, and notably in the United States, also thanks to the rising share of inherently less carbon-intensive scrap-based steelmaking. DRI aside, EAF steelmaking (increasingly in mini-mills) is the only large-scale commercial option to eliminate the use of coke, but its extent is obviously limited by scrap availability and price. When compared to a typical integrated mill, the energy requirement of a scrap-based mini-mill is just 50% (11 GJ/t vs. 22 GJ/t), carbon emissions are as little as one-quarter (0.5 t CO_2/t vs. 2.0 t CO_2/t), and the total material flux is less than one-tenth as large (0.25 t/t vs. 2.8–3.0 t/t).

Expansion of EAF steelmaking has been, despite the significant overall growth of the metal's global output, a major contributor to a relatively modest growth of industrial CO_2 emissions. Further gains of coke-free steelmaking are likely: post-2010 availability of cheap natural gas (produced by hydraulic fracturing of shales) in the United States and Canada led some experts to expect that half of North America's BF/BOF capacity will be replaced by DRI/EAF in 15 years (Laplace Conseil, 2013). Significant gains could still be achieved by near-universal adoption of the best existing practices. Given already high-energy conversion efficiencies, many specific reductions are modest, but their combination would yield improvements on the order of 10–15%, with the largest gains resulting from the installation of the best steam turbines in mill power plants, maximum use of pulverized coal injection, use of coke dry quenching, and BOF heat and gas recovery (Pardo et al., 2012).

Injection of pulverized coal has been the most successful, and now widely used, option to reduce typical coke charges. Coke dry quenching began in a few plants during the 1970s, with the pioneering installations at the NSC Yawata works able to handle 56 t/h; Japanese data show that it reached about 60% of all operations by 1990 and that it became the standard practice by 2013 (Tezuka, 2014). Red-hot (1200 °C) coke is charged into a cooling tower where its heat content is exchanged with the bottom-blown circulating inert gas, and the gas is used to generate steam in an adjacent water boiler. Most of Japan's dry-quenching plans (installed largely during the 1980s) have processing capacities of 140–200 t of coke per hour, while the largest plants in China can produce 260 t/h (NSSE, 2013).

Coke dry quenching recovers waste heat equal to about 0.55 GJ/t of coke, and, moreover, using higher quality coke made by dry quenching reduces a typical BF coke charge by 0.28 GJ/t and cuts down on dust emissions (Worrell et al., 2010). Relatively smaller energy gains would come from universal scrap preheating, sinter plant waste heat recovery, optimized sinter/pellet ratios, oxy-fuel burners in EAFs, and pulverized coal injection (Lee & Sohn, 2014). And about 10–30% of all input energy leaves EAF as hot exhaust gas, but its capture and reuse are challenging due to its high dust content. Estimated costs of CO_2 reductions range widely, depending on the targeted process, national peculiarities, and extent of controls, but they are no less than $50/t of CO_2 and could be well above $100/t.

Additional emission cuts will require new approaches, and the EU now supports a number of ultra-low CO_2 steelmaking (ULCOS) projects whose eventual aim is to cut the emissions by half (JRC, 2011). The leading techniques include top gas recycling BF and HIsarna and ULCORED processes. Top gas recycling returns the generated gas into the furnace as a reducing agent instead of preheated air, and the first demonstration plant should be ready around 2020. The HIsarna process relies on preheated coal and partial pyrolysis for melting in a cyclone and on a smelter vessel for final ore reduction, but its commercial introduction is not foreseen before 2030. ULCORED would produce directly reduced solid iron and use pure oxygen instead of air, reducing gas produced from either methane or coal syngas, and remove CO_2 by pressure swing absorption or amine washers (Knop, Hallin, & Burström, 2008). Again, the process is not expected to operate until 2030.

CHAPTER 8

Ubiquitous Uses of Steel

Sectoral Consumption and the Quest for Quality

Steel is a truly ubiquitous material: there is no industrial enterprise that would not, directly or indirectly, rely on it. There is no modern construction activity that can proceed without it: even if a building were to be constructed without any steel members or without any nails, by using doweled wood components, the wood would have to be cut by steel saws, and the building's basement would have to be dug by steel machinery, or at least by steel shovels. There is no commercial or household activity that would not, ultimately, owe its existence to it. There are no means of transportation that could function without it: even airplanes that would be made solely from aluminum and composite fibers, they would have to take off from and land on steel-reinforced runways, to say nothing about the metal used to build smelters and machines producing aluminum and carbon composites.

And this dependence only grows as societies proceed along the secular trajectory of economic development, but its forms obviously change as new uses arise and new steels are introduced to meet specific demands. Consequently, along with its enormous expansion, steel used during the past 150 years has seen major shifts in final destinations. During the 1870s and 1880s rails were the dominant product made of Bessemer steel. By 1900 the most expansive stage of worldwide railway construction was nearing its end; rails were still the leading finished product, but the twentieth century saw several major shifts in final steel uses.

At its beginning buildings with steel skeletons were still limited to a relatively small number of skyscrapers in a handful of the US cities, and the use of reinforcing steel in concrete roads, dams, and buildings was in its infancy. Car manufacturing, 8 years before launching the Ford Model T, was still a small-scale, artisanal industry akin to assembling bicycles that was producing a few thousand vehicles a year that only rich people could afford. As a result, there was still no large-scale construction of concrete roads reinforced with steel. There were no washing machines and only a small number of electric stoves; food preservation using steel cans was relatively common, but that was in part because there were still no affordable

Still the Iron Age.
DOI: http://dx.doi.org/10.1016/B978-0-12-804233-5.00008-7

electrical refrigerators. A century later the global steel output was dominated (about 55% of the total output in 2013) by hot-rolled flat pieces required to make industrial and transportation machinery and equipment and household goods, and reinforcing bars for construction accounted for nearly 15% of all shipments (WSA, 2015).

No comprehensive statistics according to consistently defined categories are kept for the metal's final consumption because steel industry's reporting ends with its output of intermediate products. World Steel Association offered the following division for 2011: construction 51.2%; mechanical machinery 14.5%; metal products 12.5%; automotive 12%; other transport 4.8%; electrical equipment 3.0%; domestic appliances 2% (WSA, 2012). Cullen et al. (2012) attempted a comprehensive and accurate mapping of the global steel supply chain for the year 2008: it traces the progress in five steps, from reduction to steelmaking, casting, rolling/forming, and fabrication.

Not surprisingly, there are many data and classification problems inherent in such complex exercises. For example, the two published values for reinforcing bar production in 2008 were 147 Mt and 210 Mt, while Cullen et al. (2012) ended up with a solved value (required to balance the total for hot-rolled long products) of 174 Mt. The final product of their analyses is a Sankey diagram, a revealing visual presentation of the scale and the complexity of steel flows. They aggregated end-use products into four major categories: vehicles, industrial equipment, construction, and metal goods. Construction was by far the largest category with 54.8% of all final steel uses, divided between buildings (32.9%) and infrastructure (21.9%). Metal goods and industrial equipment were about equal (16.3% and 16.2%), with the former split into appliances (2.8%), food packaging (just 0.7%), and other goods, and the latter divided into mechanical equipment (13.1%) and electrical equipment (3.1%). The fourth major category, labeled vehicles, claimed 12.7% of the total use, with 8% for cars, 2.8% for ships, and 2% for trucks.

McKinsey (2013) estimated the following consumption shares for China in 2015: 20% each for residential buildings, commercial buildings, and machinery, 12% for public infrastructures, and 11% for transportation, while Stahlinstitut VDEh (2014) offers the following major shares for German steel demand: 25% each for construction and cars; 13% each for machinery and pipes; and 9% for metal products. There is one thing that all of these diverse final uses have in common: as the variety of steel applications has grown, users have begun to demand higher consistence, higher quality (virtual absence of both surface and internal defects), and specific functionality, such as lighter weight, better workability, and greater strength (Iwasaki & Matsuo, 2012). The need for minimum compositional deviation led to the development of procedures that allowed mass production of steels with uniformly low contents of phosphorus (<50 ppm), sulfur (<5 ppm), hydrogen (<1.5 ppm), and carbon (<10 ppm).

New steels have filled many high-performance niches, with alloys needed for vehicles, ships, machinery, components, and tools for exacting applications ranging from mass transportation of liquids and gases to chemical syntheses and energy conversions. Their desirable physical properties include increased strength, tolerance of high temperatures, high ductility, abrasion, and corrosion resistance, easy weldability, and ability to arrest crack propagation. Better steels have also been developed for military equipment, particularly for heavy armor (steel-encased depleted uranium for tank bodies). Another notable long-term trend has been the rising demand for smaller amounts of specialty steels, including stainless steel. Besides its traditional uses for cutlery, cookware, surgical tools, and medical and food processing equipment, stainless steel has found many new uses wherever there is a need for extended corrosion resistance: in the construction of tunnels and for outdoor furniture, wastewater treatment, seawater desalination, nuclear engineering, and production of biofuels (Team Stainless, 2014).

I will look at final steel demand in some detail, focusing on three commonly used consumption categories, or subcategories (construction, vehicles, and appliances), as well as on steel used in energy systems; those uses cut across standard consumption categories as they include major infrastructural projects (such as pipelines and transmission lines), mechanical and electrical equipment (boilers, turbogenerators, transformers), and different means of mass-scale transportation (giant oil and LNG tankers, unit trains transporting coal). These surveys will be done both in global terms and by focusing on a few leading steel consumers, mostly on the United States and China.

INFRASTRUCTURES AND BUILDINGS

Certainly the most fundamental, as well as the most important (when judged by annual global consumption), use of steel products is in construction, as well as maintenance and upgrading, of buildings and key infrastructures that create housing and commercial and industrial enterprises, that underpin the functioning of modern economies, including a high degree of mobility, and that ensure public safety and private well-being. As already noted, construction of buildings and infrastructure now claims about 50–55% of annual global steel production, with national shares in low-income, rapidly modernizing economies ranging from close to, and even above, 60%, while the shares in affluent economies range mostly between 30% and 35%.

Despite the sector's importance there is no systematic reliable information on how construction uses steel. Moynihan and Allwood (2012) attempted to estimate this breakdown for the United Kingdom and on the global level. For the United Kingdom, in 2006 they concluded that infrastructure consumed 24% of all construction steel, while buildings consumed the rest, with industrial structures claiming 31% of the total. In contrast, their global allocation of 480 Mt of construction steel used in 2006 shows about 36% going into infrastructure: this expected higher share is largely explained by the urbanization and modernization in Asia. In the building sector (64% of the total), industries dominated with about 23% of all construction steel, and residential uses were only about 11%. Reinforcing steel accounts for 50–60% of all steel used in construction (Fig. 8.1);

Figure 8.1 Heavy reinforcing bars for massive pylon foundations of a bridge in Incheon, South Korea. *Corbis.*

the rest is in a variety of structural shapes, sheets and plates, rails and pipes, and tubes. As for the origins of end-use steel products in construction, the global mapping by Cullen et al. (2012) shows that about 40% of their total originated in hot- and cold-rolled coils, 28% are reinforcing bars, and about 20% become wire rods.

The list of critical infrastructures dependent on steel ranges from transportation (roads, railways, tunnels, bridges, ports, canals, airports) to environmental protection (flood and erosion control, fire suppression, air and water pollution control, solid waste disposal), and from energy systems (power plants, dams, refineries, pipelines, transmission lines) to the backbones of high-density urban living (high-rise buildings, apartment blocks, subways, industrial and commercial establishments, parks, sport facilities). Building these infrastructures *de novo* is invariably associated with the largest increase of steel consumption. But there is a fundamental difference between the past episodes of infrastructural expansion and their modern version.

In the past, as exemplified by the British, American, or German experience, these additions came in successive waves. Before 1900 the United States had seen rapid extension of railways and modern ports, its first pipelines, transmission lines, high-rises, and subways; after WW I came paved highways, the first airports, more subways, more skyscrapers, extensive pipelines, refineries, large dams, and power plants; and only after WW II did the country get its system of interstate highways, a large network of airports, and an intensive wave of electrification. In contrast, in China construction of all of these infrastructural necessities has been compressed into less than two generations.

In 1980, 4 years after Mao's death, China still had an impoverished look, with most of its technical advances traceable to the transfer of Soviet techniques of the 1950s (which, in turn, had their origin in the US advances of the 1930s). Large-scale infrastructural development began only during the 1990s and since that time the length of China's multilane highways has far surpassed the total of the US interstates (about 112,00 km vs. about 77,000 km). The country has about 16,000 km of high-speed trains (more than the rest of the world combined), and, perhaps the most telling example of the frenzied pace of its infrastructure creation, it has produced more cement (4.9 Gt) to emplace new concrete in just 3 years, between 2008 and 2010 (NBS, 2013), than did the United States (4.56 Gt) during the entire twentieth century (Smil, 2013)! And the total for the first 3 years of the second decade of the twenty-first century was even higher (USGS, 2014).

This is impressive—but also worrisome because the rapid pace of building Chinese infrastructures has not always gone hand in hand with the best achievable quality, and this will create higher-than-usual demand for maintenance and, for some structures in just two or three decades, for upgrading or replacement. Keeping all of these infrastructures safe is predicated on adequate inspection and replacement of their steel components or on their complete periodic reconstruction requiring better new steel and more advanced steel-based designs, and abundant evidence shows that even the world's richest countries have mounting deficiencies in this respect. These investment deficits have been comprehensively documented by bi-annual reports by the American Society of Civil Engineers (ASCE) that grade all infrastructures divided into four categories: water and environment, transportation, public facilities, and energy.

The latest report card awards D+ as the overall average, with solid waste disposal as the only B− grade and the rest being various Cs and Ds (ASCE, 2015). Among the six most steel-intensive categories that also require relatively frequent upkeep, only bridges and rails got a less-than-humiliating C+, while energy got a D+ and roads, transit, and wastewater treatment got a barely passable D. But a closer look at the state of American bridges makes that C+ a rather generous award. The Federal Highway Administration classifies more than 25% of the country's almost 600,000 bridges as either functionally obsolete or structurally deficient, and it estimated the investment backlog as totaling $121 billion, while ASCE (2015) calculated the need for annual investment of $20.5 billion in order to eliminate the maintenance backlog by 2028. And more steel should also be used in all seismically active regions to enhance the resilience of buildings likely to be subjected to strong earthquakes, something that has been done, so far, on a large scale only in a few affluent countries, mostly in Japan and in California (Kanno et al., 2012; USGS, 2012).

Certainly the most spectacular, that is, visually most captivating, uses of steel in construction are skyscrapers, towers, and bridges. In buildings up to 25% of all steel is structural sections, up to 44% is reinforcing bars, and up to 31% is sheet products (including roofs, interior and exterior walls, and cladding), with heating and cooling ducts, fixtures, fittings, rails, shelves, and stairs being among common nonstructural steel uses (WSA, 2012). Steel use creates larger open spaces, allowing unprecedented spans, light access, and good cross-ventilation, and these qualities have been used to the maximum in designing skyscrapers.

Construction of skyscrapers has proceeded in waves, with notable New York additions coming during the 1930s: the Chrysler building (319 m) in 1930 and the Empire State building (381 m) a year later (Landau & Condit, 1996). The World Trade Center's twin towers, destroyed in the 9/11 attacks, were 417 m tall. The tallest US structure, the Sears Tower 443 m in Chicago, was finished in 1973, and in 2015 the four tallest buildings (including spires) in the world were Burj Khalīfa in Dubai (828 m), Makkah Clock Royal Tower Hotel in Saudi Arabia (601 m), One World Trade Center in New York (541.3 m), and Taipei 101 (508 m) in Taiwan (SkyscraperPage.com, 2015).

Except for the Makkah structure, which is a steel-and-concrete complex of buildings with a central tower, all of the world's tallest 12 buildings are variations on a fundamental structural theme, with steel skeletons covered mostly by walls of glass, also with some aluminum and steel—and a large number of similarly tall, and even taller, buildings are under construction or in planning stages. These efforts have required not only high-quality, high-strength steels but also fireproof products and construction methods that guarantee flawless welding and bolting, as well as appropriate measures to limit the lateral motion of these tall structures, a goal achieved by installation of tuned dampers.

As for the world's tallest broadcasting tower, the latest record-setter is Tokyo's Sky Tree completed in 2012, an ingenious 634-m tall structure inspired by the form of an ancient Chinese tripod kettle that is built of circular pipes, with the largest bottom trusses having a diameter of 2.3 m and thickness of 10 cm, while most of the tower's outer tubing is high-strength (780 MPa) steel pipe 110 cm in diameter with a thickness of 2.5 cm (Keii et al., 2010; Fig. 8.2). The second tallest tower, in Guangzhou, is also a steel structure, in the form of a hyperboloid (twisted shape), 600 m tall, while the third highest freestanding structure, Toronto's 553.3-m tall CN Tower, is built of reinforced concrete.

Bridges have been another class of steel structure with steadily advancing record measures, both for spans and overall lengths. The longest pre–WW II suspension span was San Francisco's famous Golden Gate Bridge completed in 1937 (1280 m). By 2015 there were 11 suspension bridges with longer spans, with Japan's Akashi Kaikyo (linking Honshu and Awaji island by both road and railway, completed in 1998) holding the record with 1991 m, and cabling strength of 1800 MPa (HSBEC, 2015; Kanno et al., 2012; Fig. 8.3). The world's three longest cable-stayed bridges (in Vladivostok, Suzhou, and Hong Kong) have spans in excess of 1000 m,

Figure 8.2 Tōkyō Tree. *Corbis.*

Figure 8.3 Akashi Kaikyō Bridge linking Honshū to Awaji Island in Japan. *Corbis.*

while Ikitsuki bridge (since 1991 in Nagasaki prefecture) has the world's longest (400 m) continuous steel truss span.

In contrast to these highly visible structural sections, construction steel has also been increasingly used for pile foundations—with hat-type sectioned steel piles having end-bearing capacity as good as a steel pipe pile (Kanno et al., 2012)—as well as for steel studs (bearing and nonbearing) for external and interior partition walls of residential, commercial, and industrial buildings. Cold-formed C-shaped studs and U-shaped tracks are the most common profiles; studs are connected to track flanges with screws and in American houses are spaced at 24 inches rather than at 16 inches as with wooden studs (Steel Framing Alliance, 2007). Framing steel studs used to have a thickness of 1.2 mm; now, with stronger steels, they are just 0.6 mm thick.

FUELS AND ELECTRICITY

Modern civilization depends on incessant flows of affordable fuels and electricity—and not only is the entire energy system heavily dependent on steel, but this dependence has been increasing in both quantitative and qualitative terms, and none of the accomplishments of our advanced energy supply would have been possible without increased supply of higher-quality steels (Ginley & Cahen, 2012; Uemori et al., 2012). Laplace Conseil (2013) estimated the global steel consumption in energy sectors at 178 Mt or 12% of the total sales of finished steel products, with oil and gas production (45 Mt) and transportation (55 Mt) as the largest markets.

Large-scale underground coal mining became possible only thanks to mechanization using steel cutters and loaders, and the most efficient modern long-wall technique uses large drum-shaped cutting steel heads to extract coal from the length of a seam and dump it onto conveyors, with the miners protected under a jack-supported steel roof that advances along a seam as the cutting progresses (Osborne, 2013). And the most productive surface mining (as practiced in lignite regions of Germany or in large bituminous coal mines in the Western United States or Australia) is unthinkable without massive bucket and bucket-wheel excavators and trucks.

Extraction, transportation, storage, and use of oil and gas and refining of crude oil are no less dependent on ubiquitous uses of steel. In 1965 most hydrocarbons came from wells less than 2 km deep, offshore drilling was common only in shallow near-shore waters, the largest pipelines were the US lines from Texas to the Northeast, the largest tankers had a capacity of less than 200,000 tonnes, there were only two small liquefied natural gas (LNG) tankers. Half a century later hydrocarbon extraction commonly involves drilling wells deeper than 4 km, including many complex and long horizontal wells; offshore drilling is done in deep ocean waters (about 2.5 km); the longest pipelines carry natural gas from Northwestern Siberia all the way to Western Europe and from Turkmenistan to Xinjiang and then to Shanghai (total length of nearly 7700 km); the largest tankers carry more than 400,000 tonnes of crude oil; and the largest LNG tankers have capacities in excess of 200,000 m^3. And refineries are essentially giant assemblages of steel pipes, processing columns, and storage tanks.

Deep drilling for hydrocarbons relies on seamless steel pipe and electric resistance-welded pipe whose gastight threaded joints are coupled together to form long and heavy drill strings. These high-strength pipes must also be corrosion- and collapse-resistant as they are exposed to high temperatures, high pressures, and acidic environments deep underground and, in offshore drilling, as they hang several kilometers in seawater before reaching the ocean bottom; moreover, in directional and horizontal drilling pipes face even greater stresses as they are contorted into bends, while drilling in the Arctic exposes the rigs, pipes, and machinery to temperatures as low as −60 °C, conditions that would readily embrittle ordinary steels. Offshore extraction also requires the construction of massive drilling and production platforms whose high-strength, high-toughness steel must be particularly resistant to fracture in order to avoid catastrophic collapse of those heavy structures.

Because the volume of gas transported by a pipeline can be increased by using higher operating pressure, modern, long-distance, large-diameter trunk lines must use new kinds of ultra-high-strength pipes that also have the necessary low-temperature toughness and good field weldability. For pipelines laid at the ocean's bottom (now often in depths exceeding 2 km) collapse resistance is a key consideration, while pipelines in the regions of discontinuous permafrost or in areas affected by landslides must guard against accidental fracturing of circumferential welds and pipe buckling. And all pipelines transporting oil or gas with traces of hydrogen sulfide must be protected against hydrogen-induced cracking.

Many advances in steels for giant oil tankers and large bulk cargo carriers originated in Japan during the decades when the country was the leader in global shipbuilding, and Japanese steelmakers have continued to improve their products even after South Korea and China emerged as the world's largest builders of oceangoing vessels. The maximum strength of steel plates for shipbuilding rose from less than 250 N/mm^2 in the 1960s to 460 N/mm^2 by 2010, and the new steels have higher crack arrestability and better abrasion and corrosion resistance (Uemori et al., 2012). Besides plates tanker construction also requires various steel shapes, pipes, and bars. LNG tanks are exposed to temperatures of −160 °C: plates of 9% Ni steel (up to 5 cm thick) have sufficient brittle fracture resistance and crack propagation arrest to form large gas containers for intercontinental shipment.

High-quality steels (advanced ferritic steel, austenitic steel, and nickel alloys) have always been essential for thermal electricity-generating plants: they are the only materials to use in supercritical steam cycles where high temperature strength must be combined with good resistance to oxidation (Alstom, 2013). Chromium steel is used for the rotors of massive steam turbogenerators, whose blades are made of stainless steel or titanium alloys, while abrasion- and corrosion-resistant steels are used in capturing particulate matter in electrostatic precipitators and to desulfurize hot flue gases. Electrical (lamination) steel is indispensable for transformers to change transmission and distribution voltages: their magnetic cores are made of stacked, cold-rolled strips of iron-silicon alloys that have high permeability and low core loss (small energy dissipation per cycle), reducing electricity losses by a third compared to older designs.

Energy infrastructure is unthinkable without steel, and transition from fossil fuels to renewable sources of energy will not eliminate that dependence. The most important way of renewable electricity generation has always been steel-intensive: large concrete dams require reinforcing steel, large-diameter penstocks leading water to turbines are made of plate steel, water turbines are cast and machined from high-quality steels, and steel dominates generators as well as towers for long-distance transmission links required to bring electricity from remote regions to major load centers. And the two leading new renewable ways of electricity generation—wind turbines and photovoltaics—are even more steel-intensive.

While hydrogeneration requires 20–30 t of steel per installed MW, the rate is close to 300 t/MW for PV (Walsh, 2011). And a typical 100-m tall tower, using bolted friction connections to support a 3.6-MW wind turbine will require about 335 t of steel (83 t/MW), and a 150-m tall tower for a 5-MW turbine will need about 875 t, or 175 t per installed MW (Gervásio et al., 2014). WSA (2012) estimates per MW for onshore turbines are 30 t for the foundation, 50 t of steel for nacelle and rotor, and 100 t for tower, and, respectively, 300 t, 50 t, and 100 t for offshore machines. Many countries have bold plans for electricity from wind, but even if it were to supply only 10% of the global demand by 2030 (forecast to be about 30 PWh), then (even when using a high average capacity factor of 35% and a large average turbine capacity of 5 MW) some 175 Mt of steel would be needed to produce that output (not counting steel needed for extensive new high-voltage transmission links). In any case, wind turbines can be dismantled at the end of their useful life (now about 20 years) and steel parts can be either remanufactured or recycled.

And although steel industry is a major source of CO_2 emissions, German calculations show that new high-performance steels used in a wide range of energy converters (including central power plants, wind turbines, lighter cars and trucks, efficient electric motors) could potentially save about six times as much CO_2 as is emitted during their production (Stahlinstitut VDEh, 2013). Steel's central importance in the quest for new electricity generation is best summarized by the recent publication of a German steel group entitled *Energiewende beginnt mit Stahl* (Energy transition begins with steel), which also reminds all uncritical green enthusiasts that the process cannot succeed without competitive electricity prices (Wirtschaftsvereinigung Stahl, 2013).

Diffusion of new renewables, above all wind and solar PV generation, will change some specific forms of steel dependence: more steel will be needed for towers for wind turbines, less steel for large coal excavators. But as many sunny and windy regions (in the US windy Great Plains from northern Texas to North Dakota, sunny Southwest) are far away from major consumption centers (in the United States the eastern and western coasts), more steel will be required for new high-voltage transmission links (for tower foundations, towers, cables, and wires) in order to improve the reliability of supply and to bring electricity from large wind and solar plants to distant cities.

And although renewable conversions now receive most of the attention, nuclear fission is also expanding: according to the expectations of the early 1970s, by now it should be the dominant way of electricity generation. Obviously, this is not the case, as its Western progress became (for a variety of reasons) virtually arrested (Smil, 2010), but Asian plans call for scores of new reactors during the coming decades, mostly in China and India. Safety considerations make nuclear generation a highly steel-intensive endeavor. Zirconium steel alloy is used as the preferred cladding for fuel elements containing fissionable uranium; reactor pressure vessels, which contain the nuclear fuel encased in rods, are made of thick steel plates that are welded together in cylindrical shapes with hemispherical caps. Neutrons from the fuel in the reactor irradiate the vessel as the reactor is operated and this embrittles the metal (USNRC, 2014). Containment structures, usually dome-shaped, are made from thick (90–150 cm) steel-reinforced concrete, and radioactive wastes are temporarily stored on site in steel drums.

TRANSPORTATION

Transportation in general, and (given the large number of passenger vehicles and trucks) the vehicular sector in particular, now ranks as the leading consumer of finished steel products whose advantages include not only their strength and durability but also UV resistance and a very high (already >80% in the sector, aiming at >90% in the future) recycling rates. According to WSA (2012) all forms of transportation now claim about 17% of the global steel consumption, and road vehicles account for about 12%; as expected, automakers consume more steel, about 20% of annual production, in affluent countries (WSA, 2012). Other major consumers of transportation steel are shipyards (particularly those building massive oil and LNG tankers, container ships and bulk cargo carriers) and manufacturers of trains and airplanes.

In order to minimize energy consumption and maximize revenue-earning payload, the latter two industries now rely mostly on lighter materials (aluminum alloys, composites, plastics) to build railway cars and airplane fuselages, wings, and tails. But up to 25% of the mass of high-speed trains is steel, mainly in the heavy bogies (axes, wheels, bearings, and electric motors), while landing gears of jetliners are made from 300 M, a high-quality durable steel alloy containing chromium, manganese, molybdenum, nickel, vanadium, and silicon and providing an excellent combination of toughness, fatigue strength, and ductility (Metal Suppliers, 2015).

The mass of steel used in automobiles has been influenced by several trends that can be traced in detail in American data. First are the changes in average curb weight of cars, with a long secular increase culminating at 1692 kg in 1975, followed by a sudden decline of average mass (precipitated by oil price rises of the 1970s), a shift to smaller vehicles (the average was 1300 kg in 1985) and then yet another period of rising curb weight brought on by a spell of low oil prices and growing ownership of SUVs that lifted the average to 1470 kg in 2004. In 2013 the average light-duty vehicle (with nearly half of them being pick-ups, SUVs, and minivans) weighed 1820 kg, 1% more than a year before. In contrast to these oscillations, average steel content of American vehicles has been declining.

As a part of their pioneering analysis, Berry and Fels (1973) calculated that in 1967 91.5% (1471 kg) of an average US vehicle mass was iron and steel. That share declined to about 87% of the total mass in 1970, to 75% in 1990, and to 68% in the year 2000 (Sullivan, 2005). As Schnatterly (2008) pointed out, establishing the steel content of an average American car is not that easy: besides the direct steel shipments to automakers there are many indirect channels in the steel supply chain, imports and exports of steel components must be accounted for, and weighted curb weight average must be calculated for an ever-changing profusion of models. His detailed account established that in 2008, when average mass of all vehicles was about 1860 kg, 65% of that was steel. By coincidence, 65% of that total was steel directly shipped to automakers, and the top three finished products were, as expected, sheets and strips (in the order of galvanized and coated, hot-rolled, and cold-rolled), followed by hot-rolled bars and tube and pipe.

By 2015 about 62% of a typical US vehicle was steel, and about 70% of that total was flat-rolled carbon steel for chassis and body panels (USDOE, 2013). Cullen et al. (2012) calculated that 61% of steel in a typical car is sheet metal but because of the relatively low yield of fabrication (only about 60%), 91 Mt of it, rather than 54 Mt, were needed in 2008 (Fig. 8.4). Steel's declining share in average vehicle mass is indisputable, but as both the total and average mass of American vehicles have been increasing (from 98 million and about 1500 kg in 1970 to about 255 million and 1820 kg in 2013), steel stock contained in operating American automobiles rose from about 130 Mt in 1970 to nearly 290 Mt in the year 2013.

As expected, the global mean of automotive steel content is somewhat smaller: WSA (2012) put it at 960 kg (steel and iron) per vehicle, with roughly a third in the body, panels, doors, and trunk, about a quarter in the drivetrain (engine, gears), and 12% in the suspension, with the rest in the wheels, fuel tank, steering, and brakes. Mild steel (tensile strength of up to 370 MPa) is used for interiors and on some exposed panels, while high strength steel (HSS) (up to 550 MPa) goes for some structural parts, including doors; steels with the highest tensile strengths form chassis parts and absorption barriers. Ultra-high steels, used for side collision panels, combine high tensile strength (980 MPa and 1180 MPa) with

Figure 8.4 Coils of hot-rolled sheet steel ready for distribution at Pohang Iron and Steel Company in Pohang, South Korea. *Corbis.*

large elongation (20–50% at 980 MPa, 15–45% at 1180 MPa) and considerable bendability (Takahashi et al., 2012). Even higher strength (up to 2000 MPa) is achieved by hot stamping (heating the steel to austenitic temperature and quenching it by forming dies).

During the early 1950s the US car industry consumed about 20% of all steel shipments, and the shares were as high as 30% for bar steel and 48% for sheets. In 1950 a typical Ford four-door sedan contained about 1.6 t of steel (nearly 60% of it was heavy sheet metal); it had more color metals (Cu, Pb, Zn) than aluminum and hardly any plastic parts. Aluminum die castings made rapid inroads during the 1950s as they began to replace cast iron and sheet metal and then were used to make parts of transmissions and engines. By the early 1960s aluminum was used to cast entire engine blocks—but by 1958 average American cars still contained only about 17 kg of plastics (Hogan, 1971). That mass had nearly tripled by

1970, and the two rounds of oil price rises (1973–1974, 1979–1980) led to the introduction of smaller and lighter cars containing less steel and more aluminum and plastics.

Steel use in modern cars faces two contradictory challenges: lighter materials and lower total mass are the key steps toward making them more fuel-efficient and lowering their CO_2 emissions—but reducing car weight might compromise the quest for higher driver and passenger safety, and enhanced crashworthiness may need heavier material and additional measures (side-bags, anti-lock brakes) that will add to overall weight (Galán et al., 2012; Takahashi et al., 2012). Studies have shown that reducing vehicle mass by 10% improves fuel economy by 6–8% (USDOE, 2013)—but they also show that mass reduction is associated with increase in fatalities and serious injuries. Lightweighting has been an essential component of the quest for the already legislated higher automotive energy efficiency—the US CAFE rates for light vehicles are to reach 54.5 mpg in 2025, double the rate in 2010—and it will be even more important for large future cuts that might be necessary to reduce CO_2 emissions.

Lightweighting has also been applied to marine transportation, to the construction of large tankers, bulk carriers, and container ships, and to a rapidly expanding fleet of cruise ships. In terms of total numbers, dry bulk fleet (carrying ores, coal, fertilizers, grain, and other bulk loose cargoes) is composed mostly of vessels smaller than 55,000 deadweight tons (dwt), but the three larger categories—Panamax vessels of 60,000–80,000 dwt, capesize vessels of 80,000–200,000 dwt, and very large bulk carriers (>200,000 dwt) account for most of the fleet's carrying capacity.

Even the largest container vessels are smaller than tankers or large bulk carriers, but their capacities have grown tremendously during their rather short history (Smil, 2010). Small container ships with capacities of less than 1000 TEU (20-foot equivalent units) correspond to bulk carriers of just 14,000 dwt, and the largest ships (carrying 18,000 TEU, 400 m long and 59 m wide) are equivalent to 165,000 dtw. Global container ship fleet numbered just over 5100 vessels in 2014, compared to nearly 17,000 bulk carriers (Fig. 8.5).

Cruising has been among the fastest growing classes of tourist activities: in 2014 nearly 22 million passengers boarded the global fleet of 410 vessels, with 24 new ships to be added in 2015 (CLIA, 2014). The size of the largest of these increasingly massive vessels is rivaling the gross rate tonnage of tankers—ships of the Oasis class displace 225,282 t, have length of 360 m and height of 72 m and 20 stories—and a hull of such a vessel

Figure 8.5 Triple steel: steel ships carrying steel containers unloaded by steel cranes in the port of Hamburg. *Corbis.*

requires about 45,000 t steel covering 525,000 m^2 (Ship Cruise, 2015). Thermo-mechanical control process (TMCP) high-strength steels have been the best choice for hull construction: they require no preheating for welding and are easily formed, bent, and edged while reducing plate thickness (bottom place of the largest cruise ships are just 2 cm thick). In late 2014 ArcelorMittal signed a new contract to supply 116,000 t of steel for the hulls and decks of three new giant cruise ships, one of the Oasis class and two of the new 315-m long Vista class, to be built by STX France (ArcelorMittal, 2014).

INDUSTRIAL EQUIPMENT AND CONSUMER PRODUCTS

Both of these categories are very diverse, but one attribute they share is high shares of steel in the mass of their final products. Many pieces of common industrial and commercial equipment—ranging from filing cabinets to stainless steel wine tanks, and from distillation columns to massive shipyard cranes—are nothing but steel, and steel is also the only material used to make shipping containers. Introduction (during the late 1950s) and rapid adoption (starting a decade later) of standardized steel containers transformed the worldwide delivery of nonbulk products and revolutionized not only marine transportation but also distribution of goods by trains and trucks.

Ubiquitous containers are sized to fit trucks after they get offloaded from ships and trains, and hence their width varies only between 2.55 and 2.85 m, and their most common lengths are 6.09 m (20 ft) and 12.18 m (40 ft); their empty weight is, respectively, 2.2 and 3.7 t, and their maximum loads are 21.7 and 26.8 t (Smil, 2010). There are even longer steel shipping boxes of 45, 48, and 53 ft, and hence the world total of containers is counted in terms of TEUs. In 2015 there were about 34 million TEUs (total steel mass of about 75 Mt) in service (WSC, 2015). Their importance is perhaps best attested by the fact that all but a tiny share of all the clothes we wear, all household gadgets we use, and all electronic devices we carry reached us packed in steel containers.

Much more steel is embedded in a huge range of industrial machinery, ranging from simple but extremely powerful presses (used to stamp metals) and smaller presses to extract oils to modern, and now usually computer-controlled, machining tools including lathes, drills, gear shapers, and milling, honing, hobbing, planning, and grinding machines whose mass is typically more than 95% steel. Steel is also present in large quantities in internal infrastructures of industrial enterprises, where it forms many walkways, stairs, partitions, overhead cranes and hoists, pipes, towers, supports, and above- and underground storage tanks.

Steel that most people encounter daily in their houses and apartments has been used in making a still increasing array of kitchenware, tools, and household appliances. As already noted, manufacturing of steel cutlery predates the availability of cheap Bessemer steel, and now knives, forks, spoons, and many kinds of kitchenware are made from stainless steel varieties (Team Stainless, 2014). Post-1950 mechanization of household work in high-income countries has brought mass ownership of small appliances; the most common ones in North America have been toasters, microwave ovens, grills, mixers, fryers, pressure cookers, food processors, and juicers. But substitutions in this sector have been common, with plastics (including high-temperature-resistant and nonstick surfaces), aluminum, and tempered glass made into items ranging from cheap cutlery to baking sheets and mixing bowls.

Numbers of major household appliances (white goods) are much larger than the total numbers of road vehicles in operation: an average US household now owns two cars (the nationwide mean of car registrations is actually 1.95, slightly off its peak reached a few years ago) but it has half a dozen major appliances: refrigerator, range (gas or electric), dishwasher, washing machine, clothes dryer, and air conditioner (and many families also have a freezer and a range-like barbecue). And the United States is not alone: after decades of acquisition, ownership of nearly all of these appliances (clothes dryers, still uncommon outside the United States and Canada, are the greatest exception) has reached the saturation point in nearly all affluent countries.

Typically, washing machines, refrigerators, and ranges are owned by more than 90% of all households, and the share of families that have several air conditioner units has been increasing rapidly in such countries as Malaysia, Brazil, and China and also in the cities of India. Most people underestimate the steel content of common household appliances, which averages 56% for refrigerators, 53% for washing machines, and about 32% for air conditioners (Kubo et al., 2012). But because even major (and in the United States increasingly larger) appliances are much smaller and lighter than cars and because washing machines, clothes dryers, dishwashers, refrigerators, freezers, air conditioners, and electric and natural gas stoves are mostly made of thin sheet steel—electro-galvanized sheets with a coating weight of $20\,g/m^2$ are the norm—the share of steel production claimed by their manufacturing is surprisingly small: they are ubiquitous, but making them consumes annually only about 2% of the global steel production.

Better steels for appliances now include sheets pre-painted at steelworks (eliminating degreasing and painting by appliance manufacturers) and environmentally friendly products including chromate-free electro-galvanized steels and lead-free alloy-plated sheets (Kubo et al., 2012). In rich countries new purchases are overwhelmingly just for replacement units, while most modernizing countries have a long way to go before their ownership of basic appliances will become saturated. Refrigerators and room air conditioning units have now diffused widely among richer urbanites even in some still very poor countries (big cities in India being the prime example), but ownership of all major appliances remains low in rural regions of the Indian subcontinent, and it is virtually absent in the countryside of sub-Saharan Africa.

Most people do not think of steel and electronics as a common combination—but 47% of the mass of flat-screen TVs is steel, and stainless steel is an important component of electronic devices and components ranging from desktop computers, printers, and hard disk drives to transformers, cables, and screws. And although there is very little steel in the now ubiquitous cellphones—whose mass is dominated by plastics, special glass, and small amounts of many precious and toxic metals (Cd, Pb, Hg, As, Ni, Ag, Au)—stainless steel is used for parts that require corrosion resistance and strength, such as springs, hinges, and screws (ISSF, 2015).

CHAPTER 9

Looking Back

Advances, Flows and Stocks

Since the beginning of the twenty-first century the notion of innovation and technical progress has been, both in the public perception and in the majority of technical writings, conflated with the advances of electronics and its still expanding applications in communication, business, entertainment, and scientific research. Because of this collective infatuation it is even more important to stress the fundamentals on which this electronic edifice rests: without the constantly modernizing steel industry it would have been impossible to develop modern high-energy civilization where an unprecedented share of the global population enjoys excellent quality of life, where the number of people living in poverty in low-income countries was cut by more than half since 1990, and where there is a scientific, technical, and organizational potential to extend these benefits to additional billions.

How has this been accomplished? Iron smelting has a fascinating history longer than 3000 years, and I noted many of its fundamental developments in the first two chapters of this book. Steel, too, has a long history, but the requirements of its production kept its overall mass quite limited until after 1860: the history of large-scale commercial production of steel is thus only about 150 years old when starting the count during the late 1860s with the spreading adoption of Henry Bessemer converters in the United Kingdom. But this century and a half brought many remarkable technical advances that transformed ironmaking and steelmaking into indispensable (albeit now so curiously overlooked and unappreciated) industries supplying a variety of alloys, without whose mass-scale applications there could be no affluence for the haves and no hope for the have-nots.

No less importantly, more efficient use of raw materials and considerable reduction of specific energy consumption have been integral parts of these production advances and, in turn, they have greatly lowered the industry's environmental impact: producing today's mass of iron and steel with the techniques that prevailed a century ago would be materially wasteful, energetically unaffordable, and environmentally intolerable. I will

Still the Iron Age.
DOI: http://dx.doi.org/10.1016/B978-0-12-804233-5.00009-9

review all of these technical advances by looking at key performance variables in long-term perspectives, summarizing concisely many trends that I followed in greater detail in the first seven chapters of the book and depicting them in graphs.

Afterwards I will turn to do a brief assessment of the world's steel industry during the second decade of the twenty-first century by focusing on its many contradictions: overlooked yet indispensable; successful yet imperiled; efficient yet still perceived as environmentally offensive; innovative but fundamentally dependent on aging techniques; able to supply the demand but burdened by excessive capacity; no longer a mass employer but still the basis of productive sectors that provide good wages for millions. This will be followed by a brief quantitative recapitulation of key consumption trends (absolute and relative, in per capita terms) of the past 150 years and by appraisals of their relation to economic development and quality of life.

Continuing with historical perspectives, I will close the chapter by concentrating on the most obvious aggregate physical manifestation of the world's massive production and use of steel, namely on the rising national and global stocks of the metal, including their origins and the rates of their accumulation. Only concrete (most of it actually reinforced with steel) offers a similar example of accumulation of anthropogenic materials on scales unprecedented in the history of civilization. Annual additions to global concrete stocks are an order of magnitude larger than the accumulation of in-use steel stocks, but steel stocks are incomparably more important: concrete is rather difficult and costly to recycle, while steel stocks have come to form a new, valuable anthropogenic resource that already supplies raw material for nearly a third of global steel production and whose dependence will only grow, not only because of steel scrap's rising availability but also because of the numerous environmental benefits of its reuse.

A CENTURY AND A HALF OF MODERN STEEL

Foundations of modern steelmaking were laid by the long development and technical advances of ironmaking. Primary production of iron in blast furnaces is a perfect example of a truly medieval technique (its Western European origins date to more than 600 or 700 years ago) whose principle remains the same but whose size, degree of sophistication (both in terms of design and operation), and resulting gains in productivity have enabled vastly expanded output with reduced use of inputs and with significantly lower costs. Given their fundamental importance in producing most of the civilization's dominant metal, it is no exaggeration to rank these massive assemblies among the most remarkable artifacts of modern societies (Geerdes, Toxopeus, & van der Vliet, 2009; Peacey & Davenport, 1979; Wakelin & Fruehan, 1999; Walker, 1985).

By 1850 the best British blast furnaces designed by Lowthian Bell were fueled with coke and received hot blast from larger stoves, and their increasing volume brought higher productivity and lower unit costs of pig iron. And yet these proto-modern furnaces were just toys compared to what was to come. First, the furnaces grew taller and acquired wider hearths: by 1930 the largest ones were twice as tall and had hearth diameters nearly twice as large as in 1830; then they grew stouter: their height has increased only slightly, but by 2015 the diameters of the bellies and hearths of the largest furnaces were twice as large as in 1930 (Fig. 9.1). Doubling of hearth

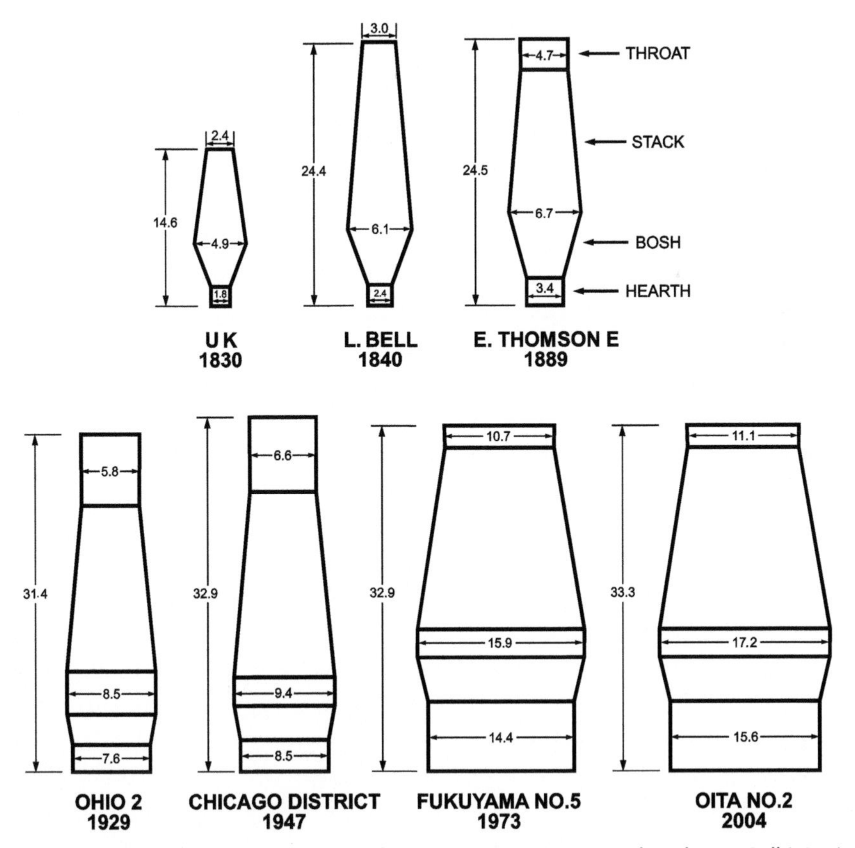

Figure 9.1 Changing designs of blast furnaces, 1830–2015. *Based on data in Bell (1884), Boylston (1936), King (1946), Sugawara et al. (1976), and Haga (2004).*

diameters results in quadrupling of hearth areas. Hearth areas increased from only about 2 m^2 in 1800 to 20 m^2 by 1910 and 70 m^2 by 1950, and in 2015 the largest furnaces had hearths on the order of 200 m^2 compared to less than 15 m^2 at the beginning of the twentieth century.

Larger hearths and taller stacks resulted in 24-fold growth of maximum inner volume between 1840 and 2015: Bell's 1840 blast furnace redesign had about 250 m^3 (compared to just 50 m^3 in 1810), by 1880 volumes of the largest furnace surpassed 500 m^3, the 1500 m^3 mark was reached by 1950, and the largest furnaces in 2015, with inner volumes of 5500–6000 m^3, were an order of magnitude more voluminous than their predecessors 100 years ago (Fig. 9.2). This dimensional growth has resulted in higher nominal capacities and rising daily productivities. The very first coke-fueled blast furnace, in 1709 in Coalbrookdale, produced just two tonnes of hot metal a day.

With the introduction of hot blast the rates surpassed 50 t/day during the 1840s, reached more than 400 t/day by the beginning of the twentieth century, approached 1000 t/day before WW II, and rose to 10,000 t/day in the mid-1970s, and the largest furnaces now produce around 15,000 t/day (and the record rate for POSCO's Pohang 4 is about 17,000 t/day), and their design capacities are close to, at, or even above 5 Mt/year, 10 times the best performances achieved immediately after WW II (Fig. 9.3). As explained in some detail in Chapters 6 and 7, these productivity advances were accompanied by declines in specific use of coke and in overall energy requirements. The earliest coke rates, during the mid-1700s (as much as 9000 kg/t of hot metal) were extremely wasteful; by the beginning of the twentieth century typical performances were 1000–1100 kg/t, and by 1950 good rates were just above 800 kg/t.

Subsequent coke rate reductions resulted from the combination of higher efficiencies of use and partial replacement by other reductants, first by oil and later by natural gas and, above all, by pulverized coal injections. By 1960 the best coke rates were well below 700 kg/t, and since the 1980s the overall consumption of reducing agents has stabilized at around 500 kg/t of hot metal, but by 2010 nationwide coke rates were about 370 kg/t in Japan and less than 340 kg/t in Germany (Lüngen, 2013). Combined with other efficiency gains (see Chapter 7), the energy cost of ironmaking has seen spectacular reductions: the earliest coke-fueled process consumed as much as 275 GJ/t in 1750; by 1900 the best rates were down to about 55 GJ/t, they were just above 30 GJ/t in 1950, and they were mostly between 12 and 15 GJ/t by 2010 (Fig. 9.4).

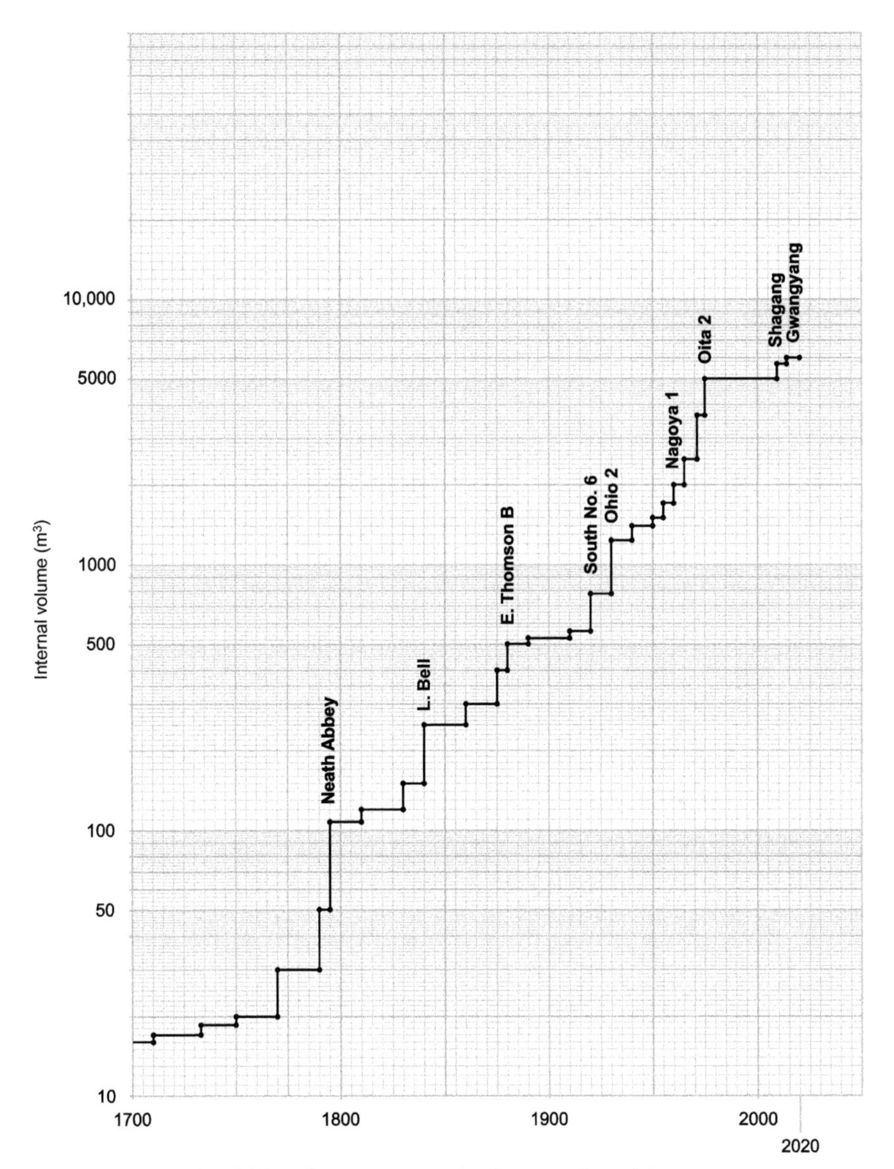

Figure 9.2 Growth of blast furnace internal volumes. *Plotted from numerous sources cited in the text.*

Large-scale commercial production of steel began with Bessemer converters; they were soon largely replaced by open hearth furnaces, and this nineteenth-century process dominated global steel output until after WW II, when it was transformed by the global adoption of basic oxygen

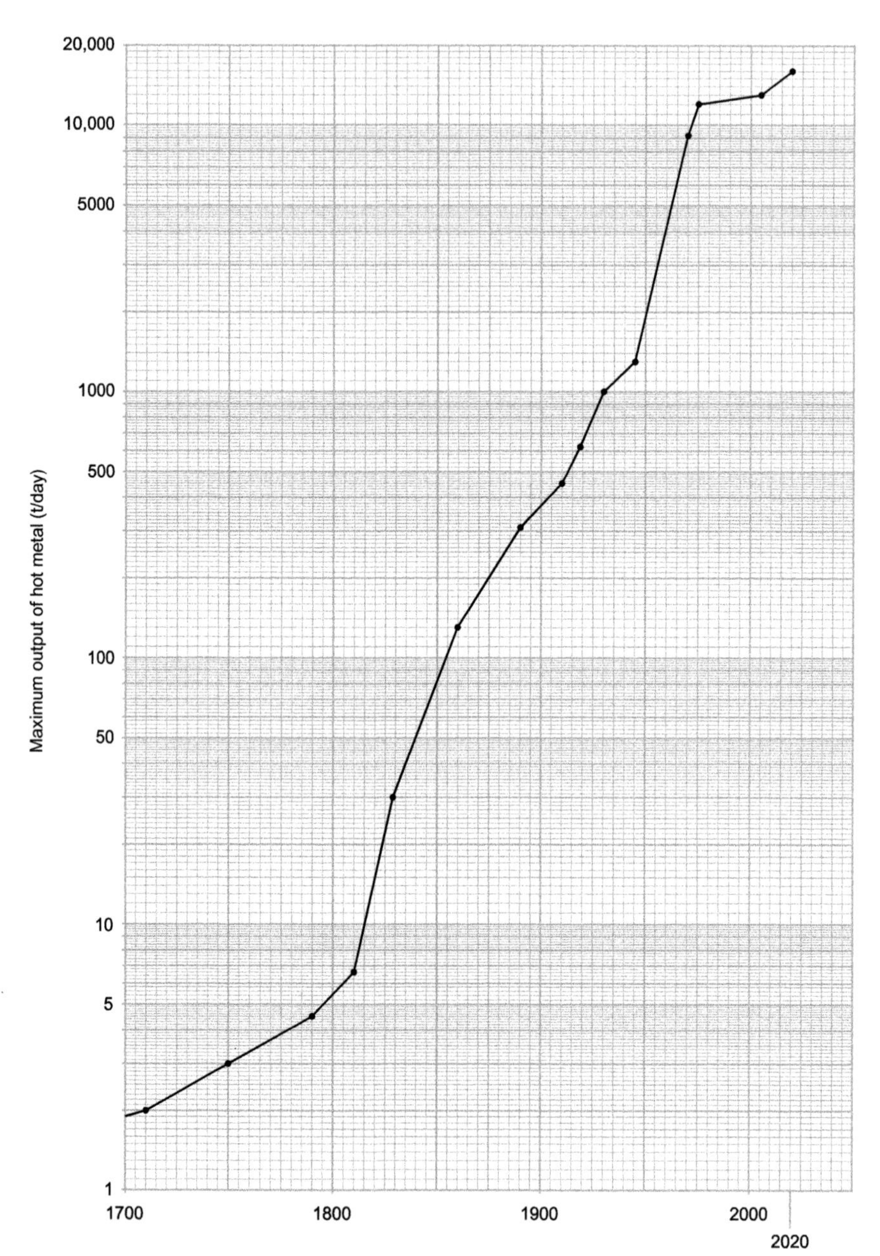

Figure 9.3 Growth of daily blast furnace production. *Plotted from numerous sources cited in the text.*

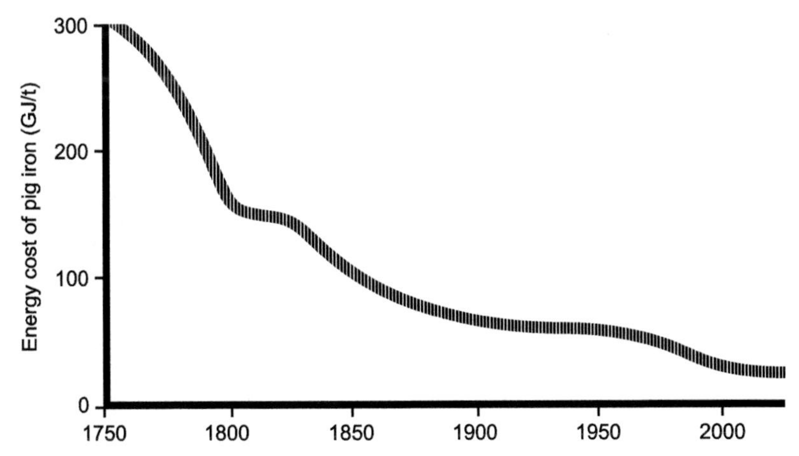

Figure 9.4 Energy cost of ironmaking, 1750–2015. *Based on Smil (2008) and on post-2000 rates cited in the text.*

furnaces and by the rising importance of scrap-based steelmaking in electric arc furnaces. One thing all of these steelmaking techniques have had in common has been the rising product yield (Takamatsu et al., 2014). When measured as a share of initial inputs, early Bessemer converters turned less than 60% of iron into steel, but eventually their yield was above 70%; early open hearths performed no better than early Bessemer converters, but a small number of remaining units now have yields of about 80%. BOFs started much higher, turning about 80% of charged iron into steel in 1952, but now that share is as high as 95%, with EAFs doing even better, with yields rising from about 85% before WW II to 97% today.

As explained in Chapter 5, major steel-producing countries differed in the onset and tempo of adopting oxygen furnaces, and the share of electric steel has been determined by the domestic availability, or the ability to import, scrap, but by 2013 those two steelmaking techniques accounted for 99.5% of total production: the tiny remainder came from technically ancient open hearth furnaces that still produced about 20% of Ukrainian steel. Globally, oxygen steelmaking went from less than half of the total in 1970 to 72% by 2013, and electric steelmaking rose from about 15% in 1970 to nearly 30% by 2015. Fig. 9.5 illustrates the complete steelmaking transition (from Bessemer convertors to OHF, BOF, and electric arc furnaces) for the United States.

Improvements in the design and operation of electric arc furnaces increased their size and productivity as they cut their tap-to-tap times,

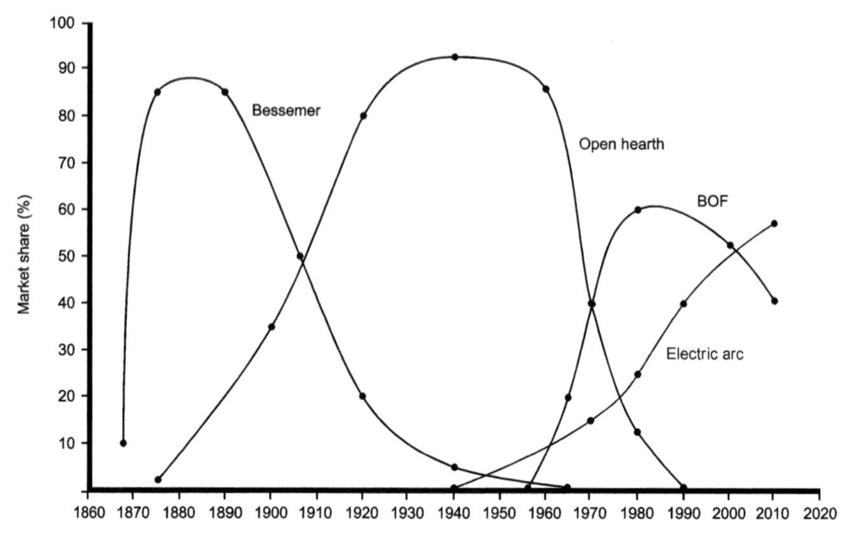

Figure 9.5 Transitions from Bessemer converters to open hearth furnaces to basic oxygen and electric arc furnaces. *Plotted from data in Campbell (1907), Temin (1964), and WSA (2014).*

electricity demand, and electrode consumption. By 2010 the shortest tap-to-tap times of about 30 min were just 1/6 of the mean in 1950, and during the same time electricity consumption for the most efficient furnaces fell by half, and electrode carbon used per unit of steel declined from 6 kg/t to just above 1 kg/t (Fig. 9.6). The last fundamental technical revolution in steelmaking was ushered in during the early 1950s by continuous casting. A classical sequence of adoption—initial slow phase (just over 5% of the total production by 1970) was followed by a takeoff (30% by 1980) and market dominance (60% by 1990), and by 2013 98.8% of the world's steel came out of continuous casting machines.

At the same time, the industry has become an impressively cleaner enterprise. Iron- and steelmaking were traditionally among the most prominent pollution-generating activities of the early and mature industrial era, a combined function of its high-energy intensity and of its lack of pollution controls. But the industry has done well during the post-1960 quest for lower energy use and reduced emissions (McKinsey, 2013; USEPA, 2012). Between 1960 and 2010 specific rates (measured per tonne of hot metal) declined by more than 40% for the overall energy consumption, by nearly 50% for CO_2 emissions (measured as weighted average of BF, BOF, and EAF production), and by 98% for dust emissions, while the accident rate was cut by about 90%.

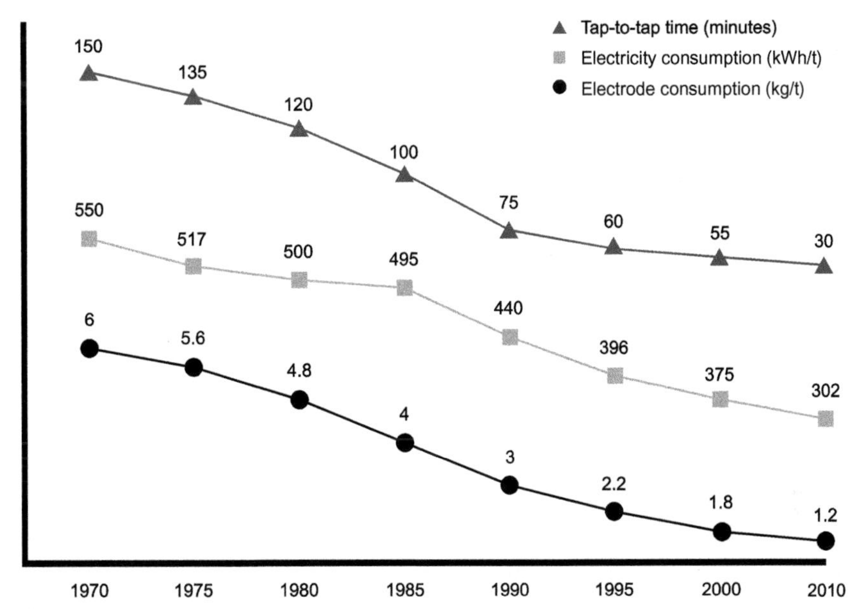

Figure 9.6 Advances in operation of electric arc furnaces. *Based on Lee and Sohn (2014).*

In 1850, before the modern steel production began, less than 100,000 t of the metal were produced annually, prorating to only about 75 g a year per capita. In 1900 global steel production reached 30 Mt, or about 18 kg/capita, while in the year 2000 the global output of 850 Mt by 2000 t translated into average annual output of 140 kg/capita, nearly 2000 times the 1850 rate. And by 2015 the rate rose further to about 225 kg/capita. Estimates of the pre-1950 world economic product are unreliable, but less questionable totals for the years 1950 and 2000 (expressed in constant $1990) indicate that the steel intensity of the global economic product was more than halved, from about 45 kg to 20 kg/$1,000, but (due to China's rapid increase of demand) by 2015 the rate was about 10% higher.

INDUSTRY'S STATE

Perspectives and judgments keep changing. At the beginning of the twentieth century no product was more fundamental to the rise of America's economic might than steel, the reality exemplified by the industry's dominance of the global market. But judging by the tenor of today's writings on technical advances and on our material world, by the beginning of the twenty-first century steel appears to be rather a *démodé* product not worthy of any close attention in the new e-world that, as the consensus has it, is dominated by silicon and lithium, by microchips, computers, portable electronics, and batteries, the New Economy that generates and accelerates its own growth.

Confirmations of these perceptions are easy to find. US Steel, the company established by Andrew Carnegie in 1901 and one of the drivers of America's twentieth-century economic dominance, was deleted from Dow Jones Industrials in May 1991, followed by Bethlehem Steel in 1997 (Dogs of the Dow, 2015). In the year 2000 the combined market capitalization of America's 10 largest steel companies had equaled just 10% of the value of Home Depot, a retail company added that year to the list of Dow industrials. For a few months during 2009 US Steel's market capitalization was below its liabilities to retiree and life insurance plans (O'Hara, 2014), on July 2, 2014, the company was removed from the S&P 500 index, and in April 2015 its market value was less than $3.5 billion, about 1.5% of what Facebook, a company that has never made anything in its short life, was worth.

And yet such economic realities are an extremely poor indicator of actual, and truly existential, importance of specific products and activities, and such irrational valuations only reflect speculative biases of investors. After all, it requires only a simple thought experiment to see how misinformed such impressions, and how unrealistic such valuations, are. Obviously, it is entirely possible to have an accomplished, advanced, and affluent civilization without any massive deployment of microchips and portable electronics—indeed, we had such a world until, respectively, the 1960s, when Intel began marketing the first microprocessor, and the 1980s, with the advent of mass ownership of personal computers and the introduction of the first cellphones. But no accomplished, advanced, and affluent civilization would be possible without mass-scale production and ubiquitous use of steel, and no investor driving up stock prices of e-companies could make it through a single day without steel.

To fulfill this critical role in the world of more than 7 billion people and in the interdependent world economy that is now worth annually more than $100 trillion, the steel industry must be large, and its enormous scale is easily demonstrated by a few global comparisons. Mining of iron ore, now surpassing 3 Gt/year, is by far the largest extraction of the metallic element, while combined primary and secondary steel production of more than 1.6 Gt is about 19 times larger than the aggregate annual output of five other leading metals, with aluminum at nearly 50 Mt, copper at close to 20 Mt, zinc at above 13 Mt, lead at nearly 6 Mt, and tin at less than 300,000 t (USGS, 2014). In addition, about 12% of total coal extraction (roughly 1 Gt) is converted to metallurgical coke, and the industry consumes every year close to 400 Mt of fluxing materials (lime, limestone, dolomite).

Given steel's key role in modern economic development, it has been unavoidable that the metal's output has reflected rather accurately the macroeconomic trends: pre–WW I expansion and WW I peak followed by sharp decline and recovery of the 1920s; crisis of the 1930s followed by rising prewar and WW II demand; post-1950 expansion checked only by the economic downturn brought by crude oil price rises of 1973–1974; then 15 years of fluctuating output before the global steel production reached a new all-time high in 1989, before a decade of slow growth or stagnation; and, most recently, the China-driven expansion that was strong enough to only slightly dent the total for 2009 (down by about 8%) and bring another record performance in 2010—and in all subsequent years until 2014.

The industry's financial footprint is much lighter because steel products are relative bargains: in 2014 output of the global steel industry was worth about $2 trillion, while the 2014 gross world product (GWP) reached about $108 trillion in terms of purchasing power parity (IMF, 2015). Consequently, steel output was worth less than 2% of the world's economic product—but that was about six times as much as the sales of semiconductors: in 2013 they reached a new record of just over $300 billion or less than 0.3% of GWP (SIA, 2014). As steel prices fell in 2015 the total worth of the industry's output decreased commensurably.

This was just the latest of recurrent changes: steel industry has always been cyclical, responding to the changes in economic activity, and the first 15 years of the twenty-first century have seen large price fluctuations. In the United States FOB prices of hot-rolled steel more than doubled between January 2000 and the summer of 2004 (from about $350/t to a peak just above $800/t), the next summer they fell below $500/t, and then they rose to a record level of $1,200/t in July 2008; a year later they were back to $400/t and have since fluctuated mostly between $600 and $800/t. At the beginning of 2015 they stood at just above $500/t and were $100 (ex-works prices) lower in Western Europe and $200 lower in China, reflecting the worldwide surplus of the metal (Steel Benchmarker, 2015).

As in every major industry, steelmaking has seen a great deal of consolidation and the emergence of large multinational companies. The largest one, Arcelor Mittal (headquartered in Luxembourg) employs more than 200,000 people in many EU countries, North America (where, among other operations, it owns the assets of the defunct Bethlehem Steel), Asia, Latin America, and Africa. South Korean POSCO (headquartered in Pohang) has plants in China, Vietnam, and Mexico; Russia's Severstal (headquartered in Cherepovets) operates also in Poland, Latvia, Ukraine, and Italy. At the same time, steel production is nowhere near as highly concentrated as it is in many other important industrial and extractive sectors.

Most notably, only two companies (Boeing and Airbus) supply the global market with large jetliners, only four companies (CFM International, GE, Rolls Royce, and Pratt & Whitney) make their engines (gas turbines), and the top 10 iron ore producers (including China's Ansteel Mining, Brazil's Vale and Samarco, and Australia's Rio Tinto, BHP Billiton, and Fortescue Metal Group) extract more than 90% of the world's supply (Mining-technology.com, 2014). In contrast, in 1990 the top 10 steelmakers produced only 18% of the world's steel, and by 2013, despite the intervening intranational mergers and the rise of new large multinational companies (ArcelorMittal operates in 20 countries), that share rose only to 27%, and the largest company, ArcelorMittal, had only 6% of the global market (WSA, 2015).

And while Boeing and GE Aerodivison have been profitable, ArcelorMittal, the world's largest steelmaker, lost $1.1 billion in 2014, $2.5 billion in 2013, and $3.8 billion in 2012 (ArcelorMittal, 2015)—and its loss-making performance has had plenty of company both among private companies and among China's state-owned enterprises (whose opaque accounting hides the real extent of their unprofitability). Between 2009 and 2013 US Steel posted 5 years of consecutive losses (O'Hara, 2014). Data collected by McKinsey show that during the first decade of the twenty-first century about 40% of large steel companies had negative cash flow (McKinsey, 2013b).

Moreover, the steel industry would require an average margin of 16% (earnings before interest, taxes, depreciation, and amortization) in order to be economically sustainable in the long term, but the mean margin between 2000 and 2014 was less than 14%, and it is not expected to recover significantly (McKinsey, 2013b). This means that the majority of steel companies cannot play an active role in the industry's consolidation and that the necessary capital will have to come from outside investors. The contrast has been stark: modern steelmaking industry is indispensable and it has become much more efficient than in the past—but it has not been sufficiently profitable, and the future of this critical industrial sector cannot be seen with assured confidence.

Massive governmental intervention (financial, legislative, directly targeted to the sector, or indirectly supportive) has been the norm for all major steel producers, and its unhappy outcomes have included deteriorating profitability, distorted markets, trade disputes, and a near-chronic excess capacity (McKinsey, 2013b). The average global capacity utilization rate was about 78% between 2005 and 2015, during the latest great economic retreat the rate dipped to just below 60% by December 2008, it was barely above 70% by the end of 2012, a year later it was still below 75%, and at the beginning of 2015 it was 72.5% (WSA, 2015). This overcapacity exists in the EU, Russia, Ukraine, and Japan, but it has been particularly troublesome in China, leading to repeated calls for the industry's restructuring and closure of up to 80 Mt of capacity by 2018, but this has been resisted by provinces and cities where steelmaking is a leading source of employment.

Although highly mechanized and equipped with numerous electronic process controls, the global steel industry still directly employs more than 2 million people (and twice as many indirectly), and that total should be doubled again by including all related suppliers. Its importance remains critical even in the countries that pioneered its modern development. Perhaps most notably, Germany—with the United Kingdom and the United States one of the three great industrial powers of the pre–WW I world—remains a manufacturing leader with steady export surpluses. Schmidt and Döhrn (2014) see steel as an indispensable cornerstone of German industry, and there are readily available statistics to confirm this judgment. Stahlinstitut VDEh (2014) estimated that in 2013 about 4 million German jobs were in steel-intensive industries (led by carmaking, machinery, and electro industries), and it calculated that in 2013 about 72% of the country's foreign trade surplus could be traced to these industries.

The two other important challenges have been the industry's aging assets (now the norm in all affluent economies) and a pronounced shift of margins to mining (a worldwide phenomenon brought by rapidly rising costs of raw materials). Steel plants in industrialized countries are getting old: in the EU 95% of all capacity was older than 25 years in 2013, in the countries of the former USSR the rate was 81%, and in the United States all of the operating plants were older than a quarter of a century (VDEh, 2013). And value creation in steel has shifted upstream, away from the steel industry to the suppliers of raw materials in general and to iron ore in particular.

In 1995 81% of profits came from steelmaking, 11% from coking coal, and 8% from the mining of iron ore, but by 2011 the shares were 26% for steelmaking, 28% for coke, and an astonishing 46% for the extraction of iron ore (McKinsey, 2013). In the year 2000 the coking coal and iron ore needed to make a tonne of hot-rolled coil were about $80 and by 2010 they surpassed $400 (McKinsey, 2010), but by 2014 the end of the expansive stage of Chinese ironmaking reduced the price of imported ore by nearly two-thirds compared to its peak in early 2011, and coking coal prices were more than halved, restoring a more acceptable division among the costs of steelmaking and its raw materials. In the closing chapter I will appraise the major factors that will either promote or counteract the reasons behind the current shifts in the global steel industry.

FLOWS AND CONSUMPTION RATES

Given the magnitude of material flows involved in the global iron and steel industry, a variety of finished products containing steel, their very different lifetimes, and the increasing importance on using the anthropogenic stores of the metal, it would be not only interesting but also quite useful to have a fairly reliable understanding of the movement and stocks of steel, inputs required to make it, and by-products of its smelting. Obviously, producing such flow-and-stock analyses is easier on a national basis and for the countries with good, long-term statistics, but (given the globalization of resource trade and the high level of manufactured exports) global appraisals would also be welcome.

Studies of anthropogenic cycles of elements in general and metals in particular have become more common since the year 2000 (Chen & Graedel, 2012)—but, as with all such large-scale, and relatively complex, flow-and-stock assessments, the results will have significant margins of error. Many flows are closely monitored (e.g., production and trade of iron ore, production of pig iron and crude steel, recovery of commercial scrap metal, steel use in automotive sector) and others can be estimated with a fair degree of accuracy (generation of iron-containing blast-furnace slag), but quantification of some flows (particularly the generation of obsolete scrap and the mass of metal abandoned in landfills or simply thrown away) requires more detailed (and often missing) information about steel content of structures and consumer products, and assumption regarding their average durability.

As noted previously, steel consumption is an order of magnitude (nearly 20-fold) larger than the combined use of the other four leading metals—but annual iron ore extraction and steel use amount to a small fraction of global material consumption whose three most massive components are minerals used in construction (sand, crushed and dimensional rock, clays), fossil fuels, and biomass (Smil, 2013). Recent annual extraction of more than 3 Gt of iron ore is surpassed by coal mining and oil production—respectively, 7.8 Gt and more than 4 Gt, with natural gas flow remaining below 2.5 Gt/year (BP, 2015)—and the global consumption of bulk construction materials, sand, clay, gravel, and stone (used in natural form or processed to make cut stone, bricks, tiles, and, above all, cement) is in the order of 40 Gt.

Starting with the flows that are known with high reliability, the global annual output of steel passed the 1 Mt mark in 1875, and by 1900 it was just above 28 Mt; the WW I peak in 1917 was 82 Mt, the pre–WW II record was reached in 1939 at 137.1 Mt, and the long post–WW II expansion peaked in 1974 at 708 Mt. After a quarter century of stagnation and slow growth the century ended with an annual output of 850 Mt, and just 2 years later began the China-led period of extraordinary expansion that brought the output to 1.43 Gt by 2010 and to nearly 1.7 Gt by 2015. These comprehensive (and fairly reliable) production statistics make it easy to calculate the aggregate global steel output. During the second half of the nineteenth century it amounted to about 400 Mt; during the twentieth century nearly 31 Gt of steel were made, with half of it after 1980; during the first 10 years of the twenty-first century the global steel industry added 11.6 Gt; and the 5-year aggregate between 2011 and 2015 was almost 8 Gt.

During the twentieth century, when the worldwide output of steel rose 30-fold, average annual exponential growth was 3.4%, but 80% of that gain came after 1950, when the annual exponential growth averaged 3%. And, largely because of China's extraordinary production increase, global output of crude steel nearly doubled during the first 15 years of the twenty-first century, to about 1.65 Gt, with the annual exponential growth averaging 4.4%. Of course, the twentieth century was also an unprecedented period of worldwide population growth, with the total rising from 1.65 billion in 1900 to 6.07 billion in the year 2000 (a nearly 3.7-fold increase and average annual exponential growth rate of 1.3%) and to 7.3 billion in 2015. This means that during the twentieth century average per capita supply of crude steel increased from just 17 kg to 140 kg (8.2-fold increase), and that it reached 226 kg in 2015 (1.6-fold increase in 15 years).

Obviously, this global average conceals enormous national disparities because steel consumption is closely associated with the rapidly ascending phase of economic growth caused by large-scale industrialization and urbanization and with associated expansion of infrastructures and rising ownership of a greater range of consumer products. During the twentieth century these processes had run their course only for a minority of the world's fast-growing population (concentrated in affluent countries of Europe and North America and in Japan, Australia, New Zealand, South Korea, and Taiwan). China, the world's most populous nation, has unequivocally joined the trend (following decades of Maoist mismanagement) since the 1990s, and India, the second most populous nation, has been at least two decades behind China.

As a result, there has been no obvious long-term link between the growth of average global per capita steel consumption and average per capita economic product: the latter rate (expressed in constant, inflation-adjusted, monies) rose 4.8 times during the twentieth century compared to a 8.2-fold increase of per capita steel consumption—but since the year 2000 the two multiples (each at 1.6) have been virtually identical. In contrast, national per capita rates generally confirm the expected association between the consumption level of steel and GDP, and some notable departures from these expectations have obvious explanations.

Historically, steel consumption can be used as a close proxy of economic overall accomplishments: average annual per capita output of steel is an excellent surrogate measure of economic advancement, and tracing this rate informs us about the fortunes of major economies and about the distance between the modal use and the leading means. In 1913, when the global mean was only about 40 kg/capita, the United States averaged about 255 kg, Germany's mean was nearly 265 kg, and the United Kingdom's rate was less than 190 kg/capita, clear indicators of the global economic leadership achieved by those three nations. A century later the global mean of crude steel consumption rose to 236 kg/capita; the US rate (at about 334 kg) was about 40% higher, and the German rate was 507 kg, but the British mean was just 153 kg/capita (WSA, 2014).

National means of affluent countries ranged between 153 kg for the United Kingdom and 1105 kg for South Korea, while China's average consumption reached about 570 kg (nearly four times the British mean), India, with 64 kg, remained an order of magnitude behind China, and the chronic underdevelopment of sub-Saharan Africa was clearly indicated by very low steel consumption rates of 14 kg in Nigeria and less than 10 kg in Ethiopia and Zimbabwe: only the relatively developed South Africa averaged above 120 kg/capita. The relationship between per capita steel consumption and per capita GDP thus shows a high correlation for low and medium rates.

A log–log plot of the national averages of the two variables shows that for GDP up to about $2000/capita and for average per capita steel consumption of 150–200 kg there is a relatively narrow scatter along a straight diagonal—but the link breaks down as countries get richer, and the plot shows considerable scatter for middle-income and high-income countries: their steel per capita consumption can range from as little as 120 kg to more than 1000 kg. Consequently, we can find countries with nearly identical steel consumption but with very different per capita GDP: both Iran and France now consume about 220 kg of finished steel products per capita but their 2013 GDPs (expressed in purchasing power parities) were, respectively, $16,000 and $40,000. Conversely, countries with nearly identical GDP have very different per capita consumption of finished steel products: in 2013 GDPs in Japan and the United Kingdom were, respectively $37,000 and about $36,000, but Japan consumed per capita nearly 520 kg of steel, while the United Kingdom consumed only about 150 kg.

This finding is not surprising as it replicates many other relationships between per capita GDP and material and life-quality indicators. Above certain levels required to satisfy what is generally perceived as a high level of economic development or satisfactory quality of life, there could be significant differences explainable by peculiarities dictated by natural conditions or specific demands of economic production. Obviously, Japan and Germany, the two leading exporters of automobiles as well as heavy machinery and entire industrial (electricity-generating, petrochemical) plants, will have a higher per capita steel consumption than France or Spain, and very high steel consumption rates for Taiwan and South Korea (respectively about 950 kg and 1105 kg per capita in 2013) are anomalies caused by the extensive use of the metal in building large oceangoing vessels (oil and LNG tankers and dry bulk carriers), and in South Korea's case also automobiles and major household appliances for export.

But obvious differences would persist even if we were to account accurately for the net steel used solely for domestic infrastructural, industrial, and commercial investment and for purchases of consumer products: again, different sectoral compositions of national economies (such as China's high share of industrial production in contrast to continuing deindustrialization of many Western economies) and differences in ownership of vehicles (and their size) and appliances will matter. And a simple quantitative measure (be it apparent per capita consumption of crude steel or finished steel products) does not capture key qualitative aspects of the metal's use, namely the increasing share of higher-value products (high-strength steel, special alloys, stainless steel) that are now increasingly required for many exacting applications.

STEEL STOCKS

The world's aggregate steel output between 1850 and 2015 was about 51 Gt, and an obvious question to ask is: what has been the fate of this large mass of metal? Looking at the twentieth century I estimated that by its end a tenth of 31 Gt of steel produced between 1901 and 2000 was oxidized or destroyed in wars and industrial and construction demolitions, about a quarter was recycled, and 15% was embedded in aboveground structures and underground (or underwater), with the remainder being the accumulated steel stock of about 15 Gt, or about 2.5 t/capita, available for potential conversion to new metal (Smil, 2005). My estimate was confirmed by the most detailed attempt to quantify the global steel stock published by Hatayama et al. (2010). That analysis put the total 2005 steel stock in 42 major countries (after doubling since 1980) at 12.7 Gt, with 60% embedded in structures and about 10% in vehicles (and with US stocks at 2.7 Gt and Japan's stores at 1.1 Gt). Raising the global total by about 20% in order to account for stocks in countries not counted in the analyses would bring it to 15.2 Gt, almost identical to my estimate.

In a more recent analysis, Takamatsu et al. (2014) used worldwide statistics to calculate the changes in accumulated global steel stock since 1870. The total reached 1 Gt around 1930 and 5 Gt in 1970, and by the century's end, as the net annual addition reached 500 Mt, it was about 12.5 Gt, or about 15% less than my approximation. By 2010 global steel stocks reached 24 Gt, and a year later they stood at 25.1 Gt, with net annual addition in excess of 1 Gt. Other published estimates offer totals that are significantly higher or lower than the range of 12.5–15 Gt for the year 2000. Müller et al. (2006) put the global anthropogenic iron stocks at 25–30 Gt in the year 2000. Kozawa and Tsukihashi (2011) estimated the global steel stocks at 27.8 Gt for 2005. And Fujitsuka et al. (2013) concluded that the global in-use steel stock had doubled between 1980 and 2010, when it reached 16 Gt.

Some researchers have tried to overcome the lack of steel consumption statistics for many countries by using nighttime satellite light images to estimate steel stock in use in engineering infrastructures and buildings.

Hsu et al. (2011) analyzed first the link between steel stocks and lights for every Japanese prefecture and then they applied the results to Japan, South Korea, Taiwan, and China. Their totals were 495 Mt for Japan and 974 Mt for China. Hattori et al. (2013) extended this approach globally, and their total for 2010 was 11.3 Gt of steel (about 21% above the 2006 level), almost equally split between infrastructures (5.5 Gt) and buildings (5.8 Gt). And Hsu, Elvidge, and Matsuno (2015) ended up with a nearly identical total of about 11 Gt of global infrastructural and building steel stocks.

National totals should be more reliable, but substantial differences are common. Müller et al. (2006) put the US total at 200 Mt in 1920, 1.25 Gt in 1950, 2 Gt in the early 1960s, 3 Gt in the year 2000, and 3.2 Gt in 2004, with per capita stocks leveling off, or even slightly declining, after 1980, when they reached just over 12 t/person. At 3.2 Gt the US steel stocks in use were just 30% less than the domestic resource base and 50% more than the reserves (at 2.1 Gt) and represented the second largest iron reservoir (after the ores in the American share of the lithosphere), followed by landfills (containing about 700 Mt), tailing ponds (with roughly 600 Mt of iron), and repositories of slag from blast and other furnaces (about 100 Mt), but errors in estimating the last three totals are particularly large (up to 50%). In contrast, Sullivan (2005) estimated the total US steel stocks in use at 4.13 Gt in 2002, and Buckingham (2006) took a closer look at steel stocks in automobiles in use and put them at 217 Mt in 2001, or just 5.3% of all steel stocks in use, implying the total of 4.09 Gt in 2001. Laplace Conseil (2013) put the total "reserves" in the US scrap "mines" at nearly 3 Gt in 2010 and the EU total at about 3.2 Gt.

Estimates of national per capita steel stocks for the world's most populous modernizing countries in 2005 ranged from just around 0.1 t for Pakistan and Nigeria and 0.4 t for India to 2.2 t for China and 3.1 t for Brazil, while the stocks in leading affluent economies were 7.5 t in France, 8.5 t in the United Kingdom, 9.0 t in Germany, 10.5 t in the United States, 12.1 t in Canada, and 13.6 t in Japan (WSA, 2012). The global average was 2.7 t, and secular trends show that, as must be expected, relatively rapid additions to stocks (prevailing until the levels reach 5–8 t/capita) are followed by much slower growth as the stocks tend toward saturation plateaux whose levels indicate substantial variation depending on the country's size and economic structure, but the range of 7–14 t/capita will accommodate all but a few economically mature outliers.

China's extraordinary post-1995 growth of steel production resulted in a rapid accumulation of the nation's steel stock. America's cumulative steel production amounted to 2 Gt during the first 50 years of the twentieth century, and during its peak production decade (1966–1975) the United States added about 1.2 Gt of domestic steel to its stock. In contrast, China added 6.5 Gt during the 10 years between 2006 and 2015, accumulating the metal's stock faster than any nation in history. China became the first nation whose total in-use stock reached 5 Gt, most likely before the end of 2010, and its stock increased to more than 7 Gt by 2015.

As just noted, the best US estimate shows that the country reached no more than 4.1 Gt between 2000 and 2002, which means that with net additions of less than 700 Mt since 2002, the US in-use steel stocks were most likely still below 5 Gt by 2015. Whatever the actual nationwide totals of steel stocks in use in the United States and China may be, there is no doubt that these two nations have created the largest anthropogenic stocks of steel, new, and largely urban and industrial, metal deposits to be "mined" profitably during the coming decades as they yield an increasing share of overall steel demand in the two countries.

Some recent publications have also focused on steel stocks in major consumption sectors. Moynihan and Allwood (2012) looked at steel use in the global and British construction sector, and Hu et al. (2010) traced the incorporation of iron and steel into Chinese residential buildings, finding the expected large difference between the steel intensity of rural and urban dwellings: the former averaged only about 5 kg of steel per m^2 of living area, while the latter incorporated about 40 kg/m^2 by the year 2010. Material flow analysis by Kawahara et al. (2012) estimated the global in-use steel stock of ships at 540 Mt in 2009 and predicted that the total will reach 900 Mt by 2035, when the annual steel demand for ship construction should increase to 41 Mt.

CHAPTER 10

Looking Ahead

The Future of Iron and Steel

Future use of any material will be affected by the progress of substitutions and the advances of relative, and eventually also absolute, aggregate dematerialization on national and global scales and by the emergence and adoption of new production processes. Even as steel has conquered new markets, it has had to cede some of its market shares to competing materials, most notably to aluminum and plastics. This process will continue, and the combination of acceptable physical attributes and substitution costs—higher for both aluminum and plastics, and much higher for new composite materials—will determine how fast and how far it will go in the future. This is also an apposite place to appraise the possibilities of another substitution, the use of modern charcoal to replace coke as part of the quest to reduce CO_2 emissions from the iron and steel industry.

Given the world's still growing population and huge unmet demand for steel in low- and medium-income countries of Asia, Latin America, and Africa (associated with the transition to modern, high-energy urbanized societies), there is no early possibility of any lasting absolute dematerialization on the worldwide scale: there may be temporary declines but in the long run global steel consumption will continue to grow. At the same time, relative dematerialization (per unit mass of a final product or per unit of national GDP) will continue, and societies will be able to derive more value and enjoy higher standards of living with progressively lower inputs of steel.

All principal iron- and steelmaking methods—blast furnaces (BFs), basic oxygen furnaces (BOFs), electric arc furnaces (EAFs), continuous casting—are now technically mature, as are the processes required to produce raw materials (extraction of coal, iron ore, and fluxing materials, production of coke, sinter, and pellets). All of these techniques will see further efficiency gains and further reductions of environmental impacts, but all of these improvements will be relatively small, and new techniques will be needed to bring fundamental technical and economic departures. I will look at the options and their likely commercial success in the chapter's penultimate section.

Still the Iron Age.
DOI: http://dx.doi.org/10.1016/B978-0-12-804233-5.00010-5

Finally, I will take a brief look at future steel consumption levels, and although I will cite some published near- and longer term estimates, I will not offer my specific (and inevitably inaccurate) forecasts but instead I will review some key factors whose combination will determine the future rates of demand and production. Existing ironmaking and steelmaking processes and their continuing gradual improvements could supply all the steel needed by the middle of the twenty-first century. Beyond that time the fortunes of the world's most fundamental metal industry will greatly depend on the pace of the unfolding global warming and on the state of our efforts to transit to economies that use the Earth's resources more rationally.

SUBSTITUTIONS

In 1900, steel had virtually no competition for many exacting, durable, or heavy-duty uses for which its mass production had been originally developed. A century later this has changed with large-scale production of aluminum and of its alloys, and with reliance on other metals for some critical applications. Titanium, 45% less dense than steel but with ultimate tensile strength only 20% lower, has been favored in alloys with aluminum, molybdenum and steel that are used above all in aerospace industry (also in bicycles, crutches, golf clubs, and hip replacements). The post-1950 world has seen a spreading use of a wide variety of plastics, and the emergence of new composite materials has cut into steel demand as they have replaced many components used to make machinery, appliances, pipes, parts, and tools. Relative steel dematerialization (share of steel in total mass of specific products) has thus been an unmistakable trend in modern manufacturing, and it has resulted in absolute steel dematerialization in many mature economies with slowly growing populations—but it has yet to translate into absolute dematerialization in global terms.

Substitution is usually seen as using an entirely different material but one of the most important trends in modern steel use has been the displacement of ordinary varieties of steel by superior alloys whose attributes translate into overall reduction in weight (and hence increased efficiency of performance when higher strength steels are used in cars or ships) or in greater longevity (with new crack- and abrasion-resistant steels). This substitution, as already noted, has made a great difference in reducing the weight of cars. Shift toward high-tensile steels is shown by the Japanese trends: during the 1990s minimum tensile strengths were below 400 MPa, by the year 2000 they rose to 600 MPa, and maxima are now near or above 1000 MPa (Takahashi, 2015).

This trend will continue. The latest concept design of Ultra Light Steel Auto Body (ULSAB) uses 100% high-strength steel (HSS), of which more than 80% is advanced high-strength steel (AHSS) (Galán et al., 2012). A WorldAutoSteel (2011) study of FutureSteelVehicle (FSV) considered more than 20 AHSS grades (expected to be commercially available by 2020) whose deployment (making up 97% of all steel in a car), combined with new design and manufacturing methods, would save up to 39% of the mass and nearly 70% of total lifetime cycle greenhouse gas emissions compared to today's designs and, given aluminum's inherently higher energy intensity, do so at a fraction of the lighter metal's cost and result in lower aggregate environmental impacts. Compared to the total vehicle masses of, respectively, 1199 and 1483 kg for smaller and larger 2010 cars powered by internal combustion engines, WorldAutoSteel (2011) sees the future steel vehicles (plugged-in hybrids) weighing 990 and 1279 kg, weight reductions of, respectively, 18% and 16%.

Lightweighting

Several lighter materials are available to replace automotive steel. Magnesium steel alloys and carbon fiber and polymer composites have the lowest density of all structural materials used in vehicles, and their widespread use has the potential to reduce component weight by more than 60%—but their widespread adoption is not imminent. Magnesium's tensile yield strength is similar to that of aluminum, but the metal has lower ultimate tensile, fatigue, and creep strength and its alloys have lower modulus and hardness, are prone to exhibit low-ductility failure, create problems with corrosion and recycling, and are difficult to be formed as sheets at low temperature (USDOE, 2013). As a result, they have been used for only about 1% of average vehicle mass. Carbon fibers and polymer composites are still fairly expensive (that is why they have been used more in airplanes rather than in cars) and also highly energy intensive.

Aluminum alloys are seen as a middle ground of the lightweighting spectrum, heavier than magnesium alloys but lighter than steel; moreover, long experience with their use in airplanes and vehicles makes their applications and limitations well understood. Car parts made of aluminum alloys include not only hoods and panels but also engine blocks and other power-train components and even entire vehicle bodies. Further use of these materials is limited due to their cost, formability, premature corrosion, and multimaterial complications in joining, painting, repairing after accidents, and recycling (USDOE, 2013). That is why HSS and AHSS are the most appealing options: their relatively high density is compensated by exceptional strength and ductility, a combination that allows weight-saving designs, and their new automotive applications have been growing faster than the substitutions by aluminum or plastics (WorldAutoSteel, 2011).

All suitable materials lighter than ordinary steel offer significant weight advantages but at higher costs. Recent costs of substitutes compared to their weight advantages (with both relative numbers for ordinary steel at 100) have been 100 and 80 for plastics (but their structural use is obviously very limited), 115 and 80 for HSS, 130 and 60 for aluminum, and 570 and 50 for carbon fiber (Heuss et al., 2012). Carbon fiber has the highest potential for weight reduction in many automotive applications, but its costs (up to six times more expensive than ordinary steel) prohibit its wide use (they are much less of an obstacle in airplane construction, with Boeing 787 being the first plane using mostly carbon fiber).

But cheaper precursor materials and their more efficient processing are expected to lower the cost of automotive carbon fiber rather significantly by 2030: conservative estimates see it down by 45%; optimistic forecasts are for reductions as high as 66%, bringing it much closer to the current cost of aluminum. In terms of weight reduction for a medium-sized car, the amount of steel substituted by lightweighting may be only 250 kg when using only HSS (and steel would still make up 63% of car mass), 420 kg when deploying more aluminum and magnesium in conjunction with HSS (all steel just 21%), and 490 kg when carbon fiber would be about 36% of the total mass and steel would account for just 16% (Heuss et al., 2012). Competitiveness of aluminum relative to steel has increased due to the falling difference in prices, and in the EU the net cost for some applications is less than 1 EU/kg, clearly a very competitive price (McKinsey, 2013a).

Consequences of a more aggressive pace of steel substitutions would be enormous, with the market for ordinary automotive steel cut by 30–50% in just a decade and even more during the next generation. Production of HSS would, of course, rise, but its future would depend on the success of lowering the cost of carbon fiber. In contrast, McKinsey (2013a) sees only negligible role for carbon fiber by 2030. Comparisons of material shares for 2010 with its modeled shares for 2030 are as follows: all steel 67% and 51% (HSS 15% and 38%), aluminum 15% and 12%, magnesium unchanged at 5%, plastics 9% and 12%, and carbon fiber 0% and 0.5%.

Lightweighting was the top emerging trend identified by leading car experts in 2014, second only to efficiency in its overall importance (Prime Research, 2014). Engine downsizing was judged to be the most promising approach, with more aluminium in second place (new Ford F-150 with high Al content is seen as shifting the metal's perception from relatively exclusive to a standard high-volume material), clever material mix in third, and, surprisingly, carbon fiber in fourth, ahead of HSS, whose role was seen as already peaking; this rise has come mainly because the experts felt that the BMW i3 could change the perception of carbon fiber from an expensive and exclusive material to one that might offer an affordable, mass-market solution for weight saving.

That remains to be seen as HSS may soon offer 25% weight reduction at no extra cost compared to ordinary steel and, importantly, when compared to other light automotive materials it will retain the advantages of component formability, durability, and safety performance; and AHSS parts can be readily repaired (Baltic & Hilliard, 2013). Another lightweighting option, particularly suitable for car roofs, is to use a sandwich material with improved stiffness: two steel sheets (0.2–0.3 mm thick) cover a layer (>0.4 mm) of polyethylene and reduce the weight by nearly 40% compared to a standard steel roof (Hoffmann, 2012). Among the promising innovations is a new alloy of iron, aluminum, carbon, and nickel (which forms a shearing-resistant intermetallic compound with Al) that has the strength and lightness of titanium alloys but costs only a tenth as much (Kim et al., 2015).

The benefits of substitution cannot always be judged simply by the cost or energy requirements of more expensive and more energy-intensive material, and only a life cycle analysis allows us to make the right conclusions. An excellent example is PE International's (2012) life cycle assessment (LCA) for complete sets of forged aluminum and steel truck wheels (18 in the United States, 12 in Europe, total distances of 1 and 1.5 million km). The study found that aluminum's higher specific burden of production (energy cost of smelting aluminum is approximately 10 times as much as producing steel) is more than compensated for during the use phase as lightweight aluminum wheels make it possible to carry additional cargo or deliver improved fuel efficiency. In a world where CO_2 emissions would be a key parameter determining material preferences, aluminum would be an indisputably better choice.

But the process can go the other way: the majority of recent replacements have used lighter materials to displace steel, but, even after 150 years of advancing steel uses, there are still applications where steel could become the dominant material. A prominent example is the replacement of wooden utility poles by steel structures. North America is a continent of wooden electricity distribution poles, with a total of 185 million units, requiring about 2.5 million annual replacements. So far, about 1 million steel distribution poles have been installed by the US utilities, and an obvious question to ask is what would be the consequences of replacing a substantial share of them by galvanized steel poles?

A revealing illustration of the complexities of this material substitution has been provided by a detailed LCA of the two options (SCS Global Services, 2013). The comparative LCA looked at six impact categories and more than 40 indicators and found that in 21 cases steel poles have a significantly lower impact (a difference of 25% or more) than wooden poles, and that for 12 indicators the difference was 100% or greater. These advantages included above all reduced depletion of energy resources (by about half) and lower (about 60%) greenhouse gas emissions, absence of exposures (of humans and ecosystems) to hazardous materials (containing arsenic and chromium) used to treat wood, absence of exposure to toxic herbicides used in forest management, and elimination of habitat disturbance and biodiversity loss attributable to large-scale tree plantations in the US Southeast. In contrast, wooden poles result in lower regional acidification (33% less), lower (40% less) ground-level ozone exposure, and, of course, vastly lower depletion of lead and zinc.

Substitution opportunities vary greatly by product category, and they can go only so far: applications ranging from the hulls of large vessels to reinforcing bars in concrete, and from high-pressure boilers to durable cutlery have no prospect to see steel displaced by other materials. In contrast, steel's share of the mass of common consumer goods will continue to decline as aluminium alloys, new plastics, and new composite materials will be making greater contributions in production of electronic gadgets and tools as well as in transportation (bogies of rapid trains are steel, but their bodies and interiors are already nothing but aluminium alloys and plastics).

Substituting Coke with Charcoal

Replacing coke with metallurgical charcoal in modern BFs would be an extraordinary challenge. Coke consumption for iron ore reduction in BFs was about 600 Mt in 2013 (additional coke is used for sintering and pelletizing ores, in cupola furnaces and in nonferrous metallurgy), and even when assuming a very conservative growth rate of pig iron production it would be about 700 Mt by 2030. As already explained, coke and charcoal have very similar energy densities of about 30 GJ/t. But their specific densities differ: preferred bulk density of metallurgical charcoal is at least $0.4\,g/cm^3$, and while a widely used Brazilian charcoal from eucalyptus surpasses that with densities of $0.53–0.59\,g/cm^3$ (Pereira et al., 2012), many trees yield charcoal whose density is well below the desirable level.

Brazil entered the twenty-first century as by far the world's largest charcoal producer and as the only major consumer of metallurgical charcoal. The fuel is converted both from illegally harvested natural forests and from plantations of fast-growing eucalyptus located mostly in Minas Gerais (Peláez-Samaniegoa, 2008). No by-products are recovered, and the conversion efficiency remains no better than 25%. About 75% of all produced charcoal is destined for BFs, with typical specific requirements of about $2.9\,m^3$ (or 725 kg) of charcoal to smelt a tonne of hot metal (Ferreira, 2000). Brazilian charcoal is used in BFs whose internal volumes are an order of magnitude smaller than those of modern coke-fueled units: the largest Brazilian furnace has just $568\,m^3$ (Pfeifer, Sousa, & Silva, 2012).

The size limit is explained by the different compressive strength of the two fuels, a key quality required to support heavy charges in large furnaces. Charcoal is too fragile to support heavy burdens in BFs whose shafts are commonly taller than 10 m and as high as 30 m and whose internal volumes are in excess of $5000\,m^3$. And while charcoal furnaces have typical volumes of just $350\,m^3$, coke-fueled furnaces are now commonly larger than $3000\,m^3$. These qualitative differences preclude any simple mass-for-mass substitution of coke with charcoal in today's large BFs using coal-based fuel.

Consequently, even if wood supply posed no constraints, charcoal's inferior compressive strength would require a massive restructuring of the industry in order to produce pig iron in smaller furnaces where charcoal could support lighter burdens. The cost of such transition would include the construction of many smaller furnaces and of the establishment of extensive tree plantations and of mass harvesting operations needed to produce the requisite wood, a change inevitably resulting not only in lower smelting productivities, increased energy costs, and higher metal prices but also in significant negative environmental impact.

Brazil offers the only instance of relatively large-scale modern reliance on charcoal in ferrous metallurgy, but using its experience as the basis for further expansion of wood-based iron ore smelting would be quite problematic. During the first decade of the twenty-first century about a third of the country's pig iron output originated in small charcoal-fueled BFs located mostly in the states of Pará, Minas Gerais, and Mato Grosso do Sul (Uhlig, 2011). Up to a third of the required wood came from illegal cutting of natural (primary or secondary) forests (Monteiro, 2006), and according to Uhlig (2011) about 15% of the Amazon's deforestation has been due to the harvests of wood for charcoaling. The rest of the charcoal was made from eucalyptus trees grown in expanding plantations reaching nearly 5 Mha in 2011 (Pereira et al., 2012).

Expanding the prevailing Brazilian charcoal-making method to meet much larger global needs would be disastrous. About four-fifths of the fuel comes from small, inefficient, and polluting semicircular brick-and-mud kilns known as hot tail, *rabo quente*. Much like their coal-coking predecessors in the early industrial United States, these 2.5-m-high brick beehives are built in massed rows, they are charged with air-dried wood, lit and let to smoulder for up to a week, and after a 3-day cooling period the fuel is unloaded and transported to BFs. Greenpeace (2013) described the typical working conditions at these charcoaling operations as a kind of hazardous slave labor. Studies have shown how underpaid workers are exposed to high temperatures, dust, smoke, and uncontrolled emissions of nitrogen and sulfur oxides, benzene, methanol, phenols, naphthalene, and polycyclic aromatic hydrocarbons (Kato et al., 2005).

Enforcement of appropriate labor regulation could solve that, but these inefficient furnaces could not be used to supply vastly increased charcoal needs. Brazilian experience shows that the charge of 450 kg of carbon per kg of pig iron can be supplied by 630 kg of charcoal (Sampaio, 2005). Global output of about 1.2 Gt of pig iron and average charge of 630 kg of charcoal per tonne of pig iron would require about 750 Mt of charcoal, about 15 times as much as the fuel's recent worldwide production (FAO, 2014). With average charcoaling efficiency of 25% (Bailis et al., 2013), this charcoal-based smelting would require about 3 Gt of wood. The actually produced total would have to be at least 5% higher because long-distance transportation (inevitable given the much higher overall global demand) and handling of such a friable fuel would result in unavoidable diminution losses, especially for charcoals made from less dense wood.

Of course, these losses could be avoided by shipping roundwood to charcoaling facilities located close to concentrations of BFs, but again, this option would result in unprecedented levels of wood trade. With a 5% markup, the charcoal-based industry would need about 3.2 Gt of wood. In comparison, recent global wood harvest has been on the order of 3.5 Gm^3 or roughly 2.3 Gt (FAO, 2014)—and wood for making charcoal would thus claim at least 40% more wood than the recent worldwide harvest used for lumber, pulp, and fuel. In order to cover all those requirements, the global wood harvest would have to rise to 5.5 Gt/year, roughly a 2.4-fold increase. And while in 2012 all wood-in-rough traded amounted globally to about 70 Mt (FAO, 2014), exporting just half of the wood needed to make charcoal (about 1.6 Gt at the 2015 smelting rate) would require a nearly 23-fold expansion of such sales.

The extent of the area that would be used to harvest this wood from fast-growing tree plantations would depend on prevailing yields. With a rather high average of 15 t/ha, it would take about 210 Mha, slightly more than half of the area of the entire Amazon basin (410 Mha). But all of these calculations could be seen as too pessimistic because the necessity to replace coke and the new, mass-scale demand for charcoal would engender many innovations that could, on the one hand, lower the specific requirements, and could, on the other hand, increase the yield of cultivated tress and the efficiency of their conversion to charcoal.

If the harvest were to come only from high-yielding (25 t/ha) clones in eucalyptus plantations (Pfeiffer, Sousa, & Silva, 2012), and if all charcoaling were to be done in modern continuous retorts that can convert 35–40% of wood to nearly pure carbon (Rousset et al., 2011), then the needed area would be reduced by nearly 60% to about 125 Mha. But, in turn, that combined assumption of high yields and high conversion efficiencies may be too unrealistic as it would be difficult to sustain 25 t/ha yields everywhere in the tropics and as such large-scale cultivation would require a great deal of nontropical charcoaling with inevitably lower harvests. Although short-term experiments on small plots indicate some impressively high harvests of temperate plantation trees, it would be unrealistic to expect yields higher than 10–15 t/ha for fast-growing hybrid poplars, pines, or willows (Smil, 2015).

For the sake of completeness, I should note that there is one exceptional source of charcoal, whose quality is easily comparable to that of metallurgical coke, but whose natural availability is very limited. Babassu palm (*Orbignya martiana*) grows in northern, northeastern, and central regions of Brazil, forming extensive monospecific forests from the inland state of Goiás to the coastal Maranhão, and it produces ellipsoidal nuts about 10 cm long and 6 cm in diameter, whose very hard endocarp has exceptionally high density (Protásio et al., 2014). Charcoal produced from the babassu nut endocarp has apparent density of 1 g/cm^3 (similar to coke) and compressive stress higher than 40 MPa, an order of magnitude higher than typical tree-wood charcoal and almost three times as high as metallurgical coke used in Brazilian BFs (Emmerich & Luengo, 1996).

Another advantage is babassu charcoal's low sulfur content. This charcoal could be thus used as a direct substitute for metallurgical charcoal in BFs—but, obviously, current supply could cover only a tiny share of potential demand. Recent Brazilian production of babassu shells is only about 1.5 Mt/year (Protásio et al., 2014), and only large-scale tropical plantations could increase its availability. Regardless of the phytomass origins, worldwide smelting of more than 1 Gt of pig iron with charcoal would have enormous environmental impacts. Relying solely on natural forests would be impossible; converting large shares of tropical forests to tree plantations would obviously be a recipe for intolerably massive deforestation and soil erosion as well as a further assault on tropical biodiversity, while maintaining large areas of high-yielding temperate trees would require repeated fertilization, applications of insecticides, and often also supplementary irrigation.

How renewable would that be? Obviously, coke-base smelting taps a finite, nonrenewable source of fossil energy, but given the magnitude of the requirement and the size of existing coal resources, we could rely on this option for generations to come even as we find other ways to reduce our mobilization of fossil carbon, above all by increasing the shares of renewable generated electricity and by replacing coal combustion by natural gas in many industrial uses.

In any case, returning to reality (recall that all of this would be possible only with creating new capacities in small BFs!), there are only two direct substitutions of coal by charcoal that make sense from both the technical and environmental points of view. The first one is to substitute charcoal for a small share of coal in coke production, and Mašlejová (2013) showed by laboratory experiments using 1–5% of wooden biomass instead of volatile coal that this substitution works. She also experimented with an indirect replacement, by using charcoal instead of fine coke in sintering furnaces. Clearly, the most appealing substitution is to use charcoal instead of injected coal: this does not create any problems with burden support and yet it could replace up to 200 kg/t of the fossil reductant. Babich, Senk, and Fernandez (2010) experimented with and modeled this substitution and found that the conversion efficiency of all the tested charcoals was either better than or comparable with the use of coal dust.

Again, just in order to indicate the magnitude of such displacement, if applied to the entire global output of 1.2 Gt of pig iron, this substitution would require about 240 Mt of charcoal, a much more manageable total that could be grown on as little as 10 Mha and no more than 20 Mha of tree plantations. But all of these calculations were done just in order to reveal the magnitudes of new material flows and ignored the economics of such substitutions. Suopajärvi and Fabritius (2013) showed that while forest-rich Finland has enough wood for current users as well as for possible ironmaking, the economics of the switch would be unfavorable, and Norgate and Landberg (2009) concluded that eventual use of charcoal in iron- and steelmaking will not be competitive with fossil fuel carbon on price alone.

Partial relief could come by using other phytomass to produce what is now called biocoke, high-density and high-quality fuel that could be made from waste phytomass, including not only woody matter from plantation thinnings but also such cellulosic wastes as bagasse (sugar cane stalks after sugar extraction) and other crop residues.

DEMATERIALIZATION

Dematerialization has been commonly defined as the reduction of material used, be it per finished product (kg/kg) or per unit of economic output (kg/$), unit of power (kg/W), performance, or service delivered (e.g., mass of a computer per instructions per second), and during the past two decades many analyses have demonstrated the ubiquity of such gains (Ausubel & Wagonner, 2008; Smil, 2013). Examples of relative dematerialization abound because the process has been one of the unmistakable defining trends in modern extractive industries, in manufacturing, energy supply, transportation, and service delivery: in all of these cases we have been using a progressively smaller mass of materials to deliver the same or even better products (lighter beverage cans, scratch-resistant yet thinner glass) and higher efficiency (plastics and composite materials in cars and airplanes) at a lower cost.

I have shown throughout this book that this relative, and persistent, dematerialization has been one of the most important accomplishments of the modern iron and steel industry. Declining quantities of ore, coal, fluxing materials, and total energy have been used to produce a tonne of hot metal in BFs, and the combination of rising conversion efficiencies in BOFs and EAFs and the universal adoption of continuous casting have reduced the material and energy demands of steelmaking. This relative dematerialization can also be illustrated by tracing the steel intensity of mature economies. When using constant GDP values adjusted for inflation and expressed in constant 2009$ (BEA, 2015) and crude steel consumption totals from Kelly and Matos (2014), the steel intensity of the US economy fell from 37 kg/$ in 1929 to 33 kg/$ in 1950. In 1973 when the country's steel output peaked, it was 20 kg/$, by 1990, following the industry's retreat (see Chapter 4), it was just 9.7 kg/$, it changed little by the year 2000 (9.5 kg/$0), and by 2013 it was down to 6.7 kg/$.

And steel has been no exception as impressive examples of relative dematerialization can be provided by analyzing secular changes in consumption of other key materials (aluminum, copper, wood, cement) per unit of national GDP or per unit weight of final products, be they airplanes or buildings. Over longer periods of time (including the 1929–2013 rates cited in the previous paragraph) this decline reflects not only gradual technical advances but it is also heavily affected by different stages of economic development: the ratio rises during the period of intensive investment in basic infrastructures; it is stagnant or falling in mature economies with slowly growing and aging populations whose GDP comes largely from less material-intensive services.

Absolute Dematerialization

Dematerialization in absolute terms has been, so far, a much rarer occurrence, but steel consumption in some mature, affluent economies is actually among the infrequent examples of this kind. Because economies continue to expand, because very few countries have declining populations, and because stationary populations may be still increasing their average per capita consumption, absolute dematerialization—declining aggregate use or even the complete elimination of a particular material on a national or even global basis—has been very rare. Perhaps its two most notable examples, elimination of chlorofluorocarbons and the use of lead in household paints—have been due to legislative actions caused by environmental and health concerns.

There has been, of course, no absolute dematerialization as far as the use of steel in the global economy is concerned: in 2013 crude steel output was 2.3 times larger than in 1973. Similarly, it is obvious that all populous, rapidly modernizing economies (above all China, India, and Brazil) have seen nothing but substantial absolute long-term increases in steel consumption. Finding out if there has been any notable absolute steel dematerialization in major mature, affluent economies requires several adjustments. In 2013, steel production in those countries was, without exception, lower than it was two generations ago: 64% lower in the United Kingdom, 38%

lower in France, 36% lower in the United States, 23% lower in Germany, and 10% lower in Japan. During those 40 years apparent crude steel consumption, taking into account all exports and imports of steel mill products, has shown absolute declines in all of these countries: 61% in the UK, 42% in France, 29% in the United States, 15% in Germany and 19% in Japan.

But given the extent of international trade in products whose mass is dominated by steel (vehicles, ships, mechanical and electrical machinery), it is necessary to adjust the apparent crude steel consumption by the net value of such indirect exports and imports. As expected, in 2013 Japan and Germany were large absolute net exporters of steel embedded in the products (mainly cars and heavy machinery) they sold abroad: the Japanese net value was 20 Mt and the German total was 8.9 Mt, while the United States indirectly imported an additional 15.3 Mt of steel, and the analogical 2013 totals were 3.2 Mt for France and 3.7 Mt for the United Kingdom (WSA, 2014). When the adjustment for indirect trade is done for these countries for the year 1973, 40-year declines in absolute steel consumption within national boundaries look like this: 17% in the United States, 26% in Germany and France, about 30% in Japan, and 40% in the United Kingdom.

The conclusion is thus clear: both apparent domestic consumption totals of crude steel and values adjusted for indirectly traded steel-intensive products show substantial absolute declines. Absolute reductions in steel consumption in affluent, mature economies have resulted from the combination of the following factors: better, less massive, product redesigns; substitution of the metal with lighter or cheaper alternatives (mainly by aluminum and plastics); shifts of economic activity from steel-intensive resource extraction (especially obvious in the United Kingdom, with abandoned coal mining) and metal-based manufacturing (compare the dominance of the US carmaking or Japanese shipbuilding during the 1970s to the recent status of those diminished industries) to services; reduced need for building construction in economies with very slowly growing (Germany, UK) or even declining (Japan's case) populations; and lower material requirements needed just to maintain rather than to expand basic infrastructures.

NEW PROCESSES

Forecasting the specifics of technical innovations is a notoriously counterproductive exercise, but the exceedingly low rate of success has not had any discernible effects on the frequency of this futile effort. Hence the next example is not to point out a particularly wrong forecast: I am using it merely as an apposite specific illustration of a common problem. In 1988, *Iron Age* examined the prospects of iron- and steelmaking, and the article quoted Egil Aukrust, the technical director of LTV Steel, who saw an early end of the era of oxygen steelmaking as a new process in which BOF and EAF get together for an in-bath smelting reduction (McManus, 1988b). And the same article cited the AISI estimate that it would take 10 years before Klöckner-CRA Technologies' direct smelting process would become commercial. Neither of these have become a reality more than a quarter century later.

Today BOFs are more important than ever—in 2013 they produced 71% of all steel compared to 57% in 1988, and their total output was roughly 2.6 times higher, 1.14 Gt compared to 442 Mt (WSA, 1990, 2015)—and in 2015 there is no credible sign (not even a plausible inkling) of their early demise. And, not surprisingly, the expected direct smelting process never became commercial, while DRI processes that have been producing iron since 1988 have continued to diffuse, but their penetration has been, as already noted, much slower than initially expected: global DRI output rose from 14 Mt in 1988 (1.8% of the total) to 75.2 Mt in 2013, still only 4.7% of the total (WSA, 1990, 2015).

None of this is surprising: developing new, commercially acceptable ironmaking and steelmaking processes is a great challenge and the progress has been commensurably slow. Perhaps the best illustration of this slow progress has been the history of the HIsmelt (High Intensity) process, promoted by its proprietor as "the world's first commercial direct smelting process for making iron straight from the ore" (Rio Tinto, 2014). Its history goes back to the development of bottom-blown oxygen converters and combined steelmaking by Klöckner Werke in the early 1970s, and its key feature is the injection of coal and fine ore into the molten bath through water-cooled lances. The ensuing reduction produces liquid iron and CO; oxygen-enriched hot blast comes from a top lance and burns the generated gas.

Advantages of the HIsmelt process include direct injections of iron ore fines (hence no sinter, no pellets) and crushed coal (cokemaking is eliminated) into the smelt reduction vessel and the use of a wide variety of possible feed materials, including hematite ore fines (even those high in P), magnetite concentrate, titano-magnetite ores, noncoking coals, and steel mill wastes. Elimination of sintering, coking, and hot blast stoves should reduce capital costs as well as space requirements and simplify operations. Trials of the process began in 1981, Klöckner tested a small pilot plant between 1984 and 1990, and a demonstration project at Kwinana (Western Australia) operated with a horizontal vessel between 1993 and 1996. A new vertical smelt reduction vessel was tested between 1997 and 1999, and its success led to an international joint venture (Rio Tinto, Nucor, Mitsubishi, Shougang) to build, at Kwinana, a HIsmelt plant with an annual capacity of 800,000 t.

At that time, Goldsworthy and Gull (2002) thought that successful operation of that plant would lead to a scale-up and that a larger HIsmelt unit, or a combination of units, would be able to replace a BF. The new plant operated between 2005 and 2008 before it was closed amid the economic downturn (Rio Tinto, 2014). According to a 2011 agreement, the plant was to be dismantled and moved to India (Jindal Steel & Power in Orissa), but the deal was canceled and the plant was bought by Molong company in China and intended for operation by 2014, with a larger pilot plant to be ready for operation by 2016 (Steel Times International, 2013). In Europe, a HIsmelt reactor has been combined with a cyclone prereducer in a demonstration plant at Tata Steel Ijmuiden (Netherlands), and the process, known as HIsarna, is a part of the European ULCOS project seeking methods for low CO_2 emission steelmaking (Birat, 2010).

Reviews of emerging ironmaking processes are available in Manning and Fruehan (2001), Fruehan (2005), Harada and Tanaka (2011), and Fischedick et al. (2014)—but all of them are conspicuous for their lack of truly new alternatives as they mostly review the accomplishments and potential of existing DRI techniques. In its Technology Roadmap Research Program, the American Iron and Steel Institute defined an ideal ironmaking process as one that eliminates the need for coal and coke ovens (and hence reduces the emissions of CO_2), that is able to use low-quality iron ores, that requires lower capital investment than the combination of coking oven and BF, and that is able to produce 5000–10,000 tonnes of hot metal a day in order to support the capacity of existing steel mills (AISI, 2010). The AISI roadmap considered six potential ironmaking alternatives and concluded that three of them could greatly reduce CO_2 emissions: suspension reduction of iron ore concentrates, molten oxide electrolysis, and paired straight hearth furnace.

The first option, whose bench-scale testing has been done at the University of Utah, is now known as Novel Flash Ironmaking (AISI, 2014). This process would use fine iron oxide concentrates directly sprayed into the furnace chamber to be reduced by gaseous agents (natural gas, now so abundantly available in the United States from hydraulic fracturing of shales, syngas, hydrogen, or a combination of these gases): no pelletized or sintered products and no coke would be needed. The process would require nearly 40% less energy, and CO_2 emissions would be reduced by 96%, about 6%, and 30% with, respectively, hydrogen, natural gas, or coal when compared to BF ironmaking and could eventually replace that traditional route.

Molten oxide electrolysis produces molten iron and oxygen as electricity passes between two electrodes immersed in a molten salt that contains dissolved iron oxide (AISI, 2010). This process, obviously, replicates the well-known and massively commercial electrolytical aluminum smelting, and its eventual (as yet poorly defined) costs would be considered along with its much-reduced CO_2 emissions. The third American innovation under investigation is the paired straight hearth furnace (PSH). The furnace is charged with cold self-reducing pellets (mixture of iron oxide and coal) whose reduction produces 95% metallized pellets that can be used in EAFs. Unlike conventional rotary hearth furnaces, whose bed height is just two to three pellets, the PSH has a bed of eight pellets (or 12 cm) deep to minimize reoxidation and to allow more efficient combustion. Eventually this furnace could be coupled with an oxy-coal melter to produce hot metal for steelmaking, a combination that would cut energy use by a third and CO_2 emissions by two-thirds compared to the standard BF route (AISI, 2014).

A techno-economic evaluation of innovative steelmaking techniques concluded that the most likely scenario is that the standard BF–BOF route, as well as BF smelting combined with carbon capture and sequestration, will become unprofitable by the middle of the twenty-first century, and that a high share of renewable energy sources and high cost of carbon will make hydrogen direct reduction and electrolysis economically attractive (Fischedick et al., 2014). This conclusion is consistent with the scenarios constructed by the authors, but the assumptions they used to build them are arguable (most notably, and not surprisingly, the German team of authors assumed a mass penetration of inexpensive wind and solar electricity).

In any case, even if successfully demonstrated, all of the new ironmaking techniques would still have to make those critical shifts from demonstration to pilot plants to mass-scale production, going against a formidable target. The combination of the thermal and chemical efficiency of modern BFs and their large working volumes, high productivity, and remarkable longevity make it very difficult to come up with a mass-reduction technique of similar performance.

FUTURE REQUIREMENTS

Perhaps the best way of assessing the future of steel production is to separate what we know for certain from what we know fairly well in general terms but are unable to assess or quantify satisfactorily, and then to stress the areas of major uncertainties. We can say with confidence that any rationally conceivable increase of steel production during the first half of this century (and almost certainly also in its second half) will not be limited by the availability of primary resources. The Earth's crust contains plenty of the iron ore, fluxing materials (limestone, dolomite), and coal needed to produce coke and to be used directly as pulverized fuel for injecting into BFs.

Obviously, every expansion of resource extraction has to end and what follows depends on the endeavor's scale. Locally, a high annual output from a particulate mine (if not its actual peak output) may be followed by a rapid decline and a complete abandonment of the site, and a similar sequence can affect even an entire mining region or a specific resource extraction that was going on in a number of sites in a small country. Resource extraction has seen many of these production peaks or brief plateaux followed by precipitous retreats and creation of economically depressed towns or regions. Among the greatest reversals are the demise of British coal mining—from 130 Mt in 1980 to the closure of the last two mines in 2015—and the emergence of the United Kingdom as a major coal importer: "bringing coals to Newcastle" is now a mundane reality, with 50 Mt imported in 2014!

This development was not caused by running out of a resource but rather by the rising costs of local and regional production, by resource substitutions (North Sea oil and gas displacing British coal), and by imports of easily available and cheaper foreign supplies. For major, globally shared resources, and for the most commonly used materials there are three key questions to ask. First, are there realistic prospects of large-scale substitution or even complete displacement by equally satisfying, or even better, alternatives? Second, if a resource looks largely irreplaceable, what is the most likely future trajectory of use? Third, how do the likely future requirements compare to the best assessment of available raw materials?

In the case of steel, the first question is easily answered in negative. When looking half a century ahead, the conclusion based on our best engineering, scientific, and economic understanding must be that there is no realistic possibility that our civilization could do without steel. The scale of the global dependence on the metal is too large to be marginalized rapidly: we use about 33 times more of it than aluminum, and nearly 6 times more of it than all plastics combined. The industry's amortization spans are several decades long: blast and oxygen furnaces and continuous casters are not built to be discarded in a few years. There is no doubt that, based on the historical experience, we will use less steel per unit of economic output or per mass of a specific durable product (benefits of relative dematerialization) and that in mature economies with near-stationary or declining populations we will see significant declines of absolute steel consumption (although in the longer term this process may be reversed by a major influx of migrants and renewed population growth).

Steel will remain the most commonly used metal of modern civilization. Even on a planet with a stationary population and minimal economic growth there would be considerable demand for steel needed to maintain and upgrade the existing infrastructures whose state ranges in most high-income countries from unsatisfactory to parlous. But the global population is set to increase during this century to at least 9 billion, and there are contradictory opinions regarding its eventual stabilization before the century's end or its continuing growth after 2100. What is indisputable is that the largest share of this growth will take place in Africa and Southeast Asia where per capita in-use steel stocks are still minuscule and where developmental needs are immense.

Nearly a billion people in Asian and African countries are malnourished, more than a billion of them still have no access to electricity, and nearly twice as many do not have a proper supply of clean water and adequate sanitation. Removing these deficiencies calls for massive increases in steel consumption to generate electricity (all kinds of central and distributed plants, high-voltage transmission), to raise food production (new fertilizer factories, farm machinery, irrigation pumps and pipes, pest-proof grain storage), and to improve water supply and treatment (dams, trunk pipes, distribution pipes for in-house delivery, wastewater treatment plants, and in an increasing number of countries also new desalination capacities). And continuing urbanization, industrialization, and expanding international trade will further add to higher steel demand for decades to come. On this basis alone, it is easy to see the potential for the 2050 global steel output to be 50% above the 2015 level.

But, as with any resource, it must also be expected that the future trajectory of steel use will have to experience, and perhaps sooner rather than later, lower growth rates, and those will eventually be followed by a global production plateau—but the time of this growth deceleration and the subsequent developments are much harder to predict. The only certain conclusion (in the absence of any affordable mass-scale substitute) is that a global production peak or a brief output plateau cannot be followed by a rapid decline without triggering enormous economic and social consequences. For example, should the global extraction of iron peak in 2050 at a level 50% higher than in 2010, it could not fall to 15% or 25% of that record level in a matter of years, or within a decade, without bringing on a massive derailment of the global economy.

And "How long can our massive steel output continue?" is the most difficult question to answer as it depends on how fast we scale up, how much we scale down, how much we recycle, and how fast we are able to come up with affordable substitutes. The common practice of relating reserve and resource estimates to recent production levels provides some useful insights, but it does not allow us to come up with the numbers for absolute future limits. The latest USGS estimate puts the global iron ore reserves at 190 Gt, containing 87 Gt of iron (USGS, 2014). At the current rate of iron ore production (3.2 Gt in 2014), the R/P ratio is just 27 years. But the global iron ore R/P has a specific shortcoming: the largest output component (China's extraction at 1.5 Gt/year) refers to crude ore rather than (as do the totals for all other countries) to usable ore, and properly adjusting that national total lowers the global output to 2.4 Gt and lifts the R/P ratio to 36 years.

R/P ratios are commonly cited, but they are not particularly illuminating: investment and innovation keep transferring usable materials from the resource category to economically exploitable reserves, and hence we can be absolutely certain that there will be plenty of iron ore in 27 or 36 years. Looking at the resource to production ratio is more revealing, and the USGS puts the total iron content of global ore resources at 230 Gt: with annual output at the 2014 level of 1.6 Gt, those resources would last just over 140 years, far beyond any rational planning horizon: think of the industrial planners worrying in 1871 (just 6 years after the end of the US Civil War and during the year of France's defeat by Prussia, when the global steel output was less than 1 Mt) about the metal's production in 2015. Obviously, concerns are different at the national level. Perhaps most notably, if iron ore exports from Brazil and Australia were to continue along their rising post-2000 trajectory, their stocks could be depleted rather rapidly (Yellishetty & Mudd, 2014).

Moreover, 30% of those 1.6 Gt produced in 2014 came from recycled metal, and the primary output of 1.1 Gt would extend the life of estimated iron resources to more than two centuries: think of the Congress of Vienna in 1814 worrying about iron ore supply in 2014 rather than about the organization of post-Napoleonic Europe. McKinsey (2013a) modeled the future iron ore demand using three global scenarios: a high-demand version would see 2.69 Gt extracted in 2020 and 3.38 Gt in 2030, a steady-growth scenario would end up with 2.44 Gt in 2020 and 2.63 Gt in 2030, and an early market saturation would call for only 2.4 Gt by 2020 and 2.35 Gt by 2030.

And, as the review of flows and stocks made clear, yet another key adjustment must be made: to account for the secondary resource of the metal that is being constantly created by increasing steel stocks in buildings, infrastructures, vehicles, appliances, and other steel products, and whose increasing shares will be recycled, not only because it is conveniently available but also because its reuse reduces the industry's energy intensity and environmental impacts, above all its substantial greenhouse gas emissions. Pauliuk et al. (2013) modeled global steel demand during the entire twenty-first century (an inherently uncertain exercise) and concluded that the demand for primary steel may peak as early as the year 2025 (creating an increasing excess of BF capacity during the century's third and fourth decades); that the EAF route (including DRI) will reach parity with the BF–BOF path by 2050; that it will deliver nearly twice as much metal by the century's end than BOFs; and that in some regions the rising availability of scrap will be larger than the final steel demand.

Forecasts

As in so many other instances, the simplest forecasts of steel production or steel demand are done by extrapolating recent growth rates in major steelmaking countries. But, obviously, sudden economic downturns or unexpected output spurts can take this approach way off the mark, even in fairly short periods of time. Seventy years of post-WW II changes in global steel production show how unadvisable it would be to take even a decade (in economic affairs a moderately long time span) as a basis of long-range forecasts. Between 1945 and 1974 worldwide steel production grew steadily, its increment averaging nearly 6% a year—but that growth (much like that of virtually all important economic variables) came to an abrupt end with OPEC's quintupling of oil prices in 1973–1974. What followed was more than a quarter century of stagnation (less steel was produced in 1993 than in 1978) and slow growth as overall declines of Western production were only modestly surpassed by expansion in a few populous modernizing countries, above all in China, India, Brazil, and South Korea.

In 2001, the global output of 852 Mt was only 20% above the 1974 level of 708 Mt, implying annual exponential growth of just 0.7%. But then came the Chinese takeoff, and output stabilization and some marginal recovery in affluent countries. Average annual growth of the global steel demand, which was just 1.9% per year during the last decade of the twentieth century, jumped to 7% per year between 2000 and 2005; it was 4.4% between 2005 and 2010 and almost 3% between 2010 and 2015. A slowdown has clearly been under way, and even slower growth is most likely in the near future as the China-driven rise between 2000 and 2013 was an exception that is unlikely to be repeated. McKinsey (2014) expects an average annual growth rate of 2.8% until 2020 (extremes of 2.2–3.5%) and 1% during the 2020s (range of 0.6–1.4%), close to the pre-2000 rate. Obviously, most of this growth will continue to be dominated by Asia, with China's share of global steel output declining but still claiming 42% by 2026 and 38% by 2030, and with India still lagging but rising to about 10% by 2030.

The other common forecasting choice is to relate future steel consumption to rising GDP. But (recall those substantial differences in national steel/GDP rates) this approach has to reckon both with many idiosyncratic national consumption patterns and, moreover, it can rely on (more or less) linear extrapolation only during the early stages of economic development. Pauliuk, Wang, and Müller (2013, 22) argue correctly that in order to understand future steel demand, "a detailed understanding of the evolution of steel stocks . . . is indispensable." At the same time, forecasting future stock levels cannot be done with a high degree of accuracy as any such modeling exercise is affected by the choice of analytical boundaries, by the degree of sectoral subdivision used to trace stock levels, by the rates of scrap generation and consumption, by the extent of metal's losses in fabrication, and by the level of international trade.

Addition to stocks will, obviously, also depend on average lifetimes, and relatively small variations in assumptions can add up to substantial differences decades later. Assumptions for lifetimes of steel in construction range from 40 to 100 years, in transportation from 10 to 30 years, and in machinery and appliances from 10 to 40 years (Müller et al., 2006). Pauliuk et al. (2013) used averages of 15 years for products, 20 years for transportation, 30 years for machinery, and 75 years for construction. In another study, Pauliuk et al. (2013) used ranges of 13–27 years for transportation, 15–40 years for machinery, 38–100 years for construction, and 8–20 years for products. But Kozawa and Tsukihashi (2011) favored generally much shorter average lifetimes, just 35 years for infrastructure, 30 years for buildings, 14 years for machinery, and 9 years for cars.

Pauliuk et al. (2013) performed analysis of in-use steel stocks for 200 countries (noting large uncertainties surrounding many of their values) and found clear saturation levels for all use categories. Per capita rates in industrialized countries range from 10 ± 2 t for construction to 0.6 ± 0.2 t for appliances and other steel products, while transportation steel stocks appear to saturate at about 1.5 t/capita and machinery stocks at 1.3 t/capita, and the overall steel saturation level is 13 ± 2 t/capita. More specifically, Pauliuk et al. (2013) forecast per capita saturation levels at 13.2 t for North America by 2020, 15.4 t for rich East Asian economies also by 2020, 12.8 t for Western Europe by 2030, 13.7 t for China by 2050, and the same high rate for India and Africa but both as far in the future as 2150.

In contrast, Fujitsuka et al. (2013) estimated that the saturation value for steel used in Asian buildings is twice as high as in other areas, the rates explained by the high incidence of earthquakes in East Asia (Japan, South Korea, China) where most of the recent steel consumption has taken place. They also estimated that by 2050 the global in-use stock of steel in infrastructures and cars will be four times the 2010 level, while for buildings it will be five times the 2010 level. And, assuming that the reuse of steel will remain at the rate prevailing in the early 2010s, they concluded that by 2035 the obsolete scrap generation will become about twice as large as the demand for the metal. And even if the near-term Chinese steel production declines by as much as 20%, annual additions to steel stocks should still surpass 1 Gt by 2020, and global in-use stocks should increase by at least 10 Gt between 2015 and 2025. Not surprisingly, Hatayama et al. (2010) forecast that these increases will be driven mostly by an expected 10-fold rise of steel consumption in Asia and that the total for structural and vehicular stock will reach 55 Gt by 2050.

This is helpful to indicate the national and global levels of an eventual equilibrium, but it is of limited help in forecasting steel demand during the next 10–30 years, when the difference in economic development and national peculiarities may be the most important determinants of output and trade. Substantial steel demand will persist even in affluent and aging societies with stable, or even declining, populations. More durable designs may lengthen the recycling spans for cars and appliances, but if affluent societies are to enjoy continued high standards of living, they will have to address their deficits in infrastructural maintenance, and this would be a source of persistent steel demand. Continuing urbanization (particularly more highrises in expanding megacities) will keep creating strong demand for steel in construction and infrastructural sectors (especially for rapid transport, including steel-intensive subways) in all rapidly modernizing populous countries.

The continuing rise of car ownership—the global count of passenger vehicles had surpassed 100 million during the late 1950s (most of them still in the United States) and 1 billion in 2005, with China being the largest market since 2009 (with 23.5 million units sold in 2014)—will mask the difference between saturated, and even slightly declining, markets in rich countries and large potential growth throughout sub-Saharan Africa and among the poorest half of all populations in Asia and Latin America. As with so many other capital-intensive acquisitions, the future increases in ownership depend primarily on rising incomes, but lightweighting will also have a major impact on steel demand in the still-expanding car industry. Production will grow, but as a result of lightweighting (and also of gradual penetration of less steel-intensive electric vehicles), the sector's total steel demand may see only a marginal increase by 2025 (McKinsey, 2013b).

At the national level, it is safe to conclude that China's experience will not be repeated: between 1992 and 2000 the country accounted for 35% of the world's growth of steel production, between 2000 and 2007 that share rose to 62%, and between 2007 and 2014 it reached 98%. India is the only similarly sized country that could replicate that achievement, but its economy clearly has not taken off as rapidly as China's. So far, the slowdown in China's steel output has been gradual, but with the expansionary period over, it is natural to ask how far the maximum output will go and how soon a new production plateau will get established.

Arguments for retreat are obvious. While the unmet demand in the poorest provinces (Guizhou, Yunnan, Gansu) leaves more room for growth, the industry's overall large overcapacity, already high consumption levels in the most economically advanced provinces, unprofitability of small enterprises, and enormous environmental burdens point in the other direction. Most notably, Tianjin and Shanghai, two municipalities under the central administration, and the provinces of Jiangsu, Zhejiang, Liaoning, and Nei Monggol (with about 240 million people) already have average per capita steel use surpassing the peak US levels of the 1970s, before the American steel production began to decline.

Not surprisingly, 2015 brought the first signs of retreat after decades of advances: in January 2015, China's crude steel output was nearly 5% below the level of January 2014. But that does not mean that a secular decline has begun. Perhaps the best bottom-up analysis of China's long-term steel demand offers a plausible sequence of rising and falling output. Yin and Chen (2013) concluded that the demand will peak at 753 Mt in 2025, and then decline to 510 Mt by 2050. As for sectoral use, construction demand should fall while automotive demand should rise rapidly before 2035, and it should claim 19% of the total output by 2050 (compared to just 6% in 2010).

The other key determinant of future output will be the fate of the country's effort to improve the quality of its incredibly polluted air. In February 2014, more than 15% of China's territory experienced exceptionally heavy smog, with maximum daily levels of fine particulate matter (PM 2.5) in Beijing surpassing 900 μg/m^3, while the WHO's acceptable maximum is 25 μg/m^3 (Dai and Gutierrez, 2014). Coal-fired electricity generation is by far the largest source of China's air pollution, but steelmaking ranks as the second largest emitter of particulates and CO_2, and the worst situation is in the regions of high steel mill concentrations in northern Hebei province (especially Tangshan municipality), which produce one-third of the country's large steel output.

In contrast, Credit Suisse (2012) argued that any steel consumption peak was "highly unlikely anytime soon," mainly because China's capital stock (including in-use stocks of steel) is still very low in relative terms (an order of magnitude lower than in the United States) and because the key factors driving higher consumption (demand by construction, transportation, and infrastructural improvements) will remain for years to come, while the expected structural decline "may take far longer than most expect." Similarly, Walsh (2011, p. 36) concluded that "the China story has a long way left to run," pointing out that the country's average steel intensity surpassed 500 kg/capita only in 2011. In contrast, the United States steel intensity remained above that mark for 30 years, and by 2015 both Japanese and German intensities had been at such a high level for nearly 40 years, and South Korea for more than 30 years.

An obvious counterargument is that these three countries have been exceptional exporters of steel-intensive cars and machinery and that there seems to be no early prospect of Chinese brands conquering global markets to such an extent as Toyota, Honda, Volkswagen, Daimler, Audi, BMW, or Hyundai have done. And it is also unlikely that in the next two decades India will come close to replicating China's post-1990 rise. The combined population of the Indian subcontinent (of India, Pakistan, and Bangladesh) is already larger than that of China (1.6 billion in 2014, compared to 1.4 billion in China), and by 2025 (when their combined population will reach 1.8 billion) their steel demand will be only a third of China's 2014 rate of about 550 kg/capita. Their annual steel consumption was about 325 Mt compared to about 90 Mt in 2013.

A draft of India's new national steel policy aims at "transforming Indian steel industry into a global leader in terms of production, consumption, quality, and techno-economic efficiency," and a combination of domestic and foreign investment is planned "to reach crude steel capacity level of 300 million tonnes by 2025–26 to meet the domestic demand fully" (Government of India, 2012). And there is even greater potential demand throughout sub-Saharan Africa, with both of these regions having enormous infrastructural needs. McKinsey Global Institute (2013) put the global infrastructural spending between 2013 and 2030 at about $57 trillion, with steel-intensive projects accounting for more than 75% of the total—and while during that period India should finally realize many of its long-deferred economic aspirations, it is unlikely that either region will come close to sustaining more than a three-decade-long spell of intensive economic development akin to China's post-1980 performance.

Europe remains a major consumer of steel, but again, the continent's future steel demand may follow a course of very gradual decrease (easily explained by the presence of already dense infrastructures, saturation of car and appliance ownership, and declining fertility) or of a steeper decline (resulting from chronic economic problems, rapid population aging, and stresses caused by political instability and mass immigration) that might end only with a formation of a new, and much lower, in-use equilibrium. North American prospects are no less uncertain: on the one hand there are already very high ownership levels of vehicles, appliances, and houses that bring steel used by those sectors close to, or even above, what may be the most likely long-term saturation level; on the other hand are huge, and growing, infrastructural deficits whose remedy would require substantial investment in steel.

Not surprisingly, even the organizations seen as the most able forecasters can do no better than to resort to scenarios with widely differing outcomes. McKinsey (2013a) offered three very different scenarios of future steel demand: the high-growth version (labeled Scarcity) ends up with 1.94 Gt in 2020 and 2.62 Gt by 2030; in the moderate version (Steady Growth) the total remains below 2 Gt by 2020 (1.81 Gt) and rises to 2.21 Gt by 2030; and in the low-growth version (Saturation scenario) the total rises only modestly to 1.79 Gt in 2020 and to just above 2 Gt (2.06 Gt) by 2030. But none of these forecasts considers potential consequences of accelerated global warming.

Continuing inadequacies of our long-range models make it impossible to offer confident predictions of the future global temperature rise (Palmer, 2014). As a result, we cannot exclude the possibility that, although not highly probable, some of the more extreme scenarios (e.g., average global temperature rise of more than 3 °C by 2050) may become a new reality. Obviously, such an unprecedented warming would call for unprecedented steps to manage the challenge—and steel production, the source of nearly 10% of all CO_2 emissions from the combustion of fossil fuels, would be affected by efforts aimed at lowering the annual additions of carbon to the atmosphere. If that were the case, the third century of mass steel production, that will begin during the 2060s, would be very different from the first one that provided the foundations of modern civilization, and from the second one that has been extending its benefits worldwide.

APPENDIX A: UNITS AND THEIR MULTIPLES AND SUBMULTIPLES

BASIC SI UNITS

Quantity	Name	Symbol
Length	meter	m
Mass	kilogram	kg
Time	second	s
Electric current	ampere	A
Temperature	kelvin	K
Amount of substance	mole	mol
Luminous intensity	candela	cd

OTHER UNITS USED IN THE TEXT

Quantity	Name	Symbol
Area	hectare	ha
	square meter	m^2
Electric potential	volt	V
Energy	joule	J
Force	newton	N
Mass	gram	g
	tonne	t
Power	watt	W
Pressure	pascal	Pa
Temperature	degree Celsius	°C
Volume	cubic meter	m^3

MULTIPLES USED IN THE SI

Prefix	Abbreviation	Scientific Notation
deka	da	10^1
hecto	h	10^2
kilo	k	10^3
mega	m	10^6
giga	G	10^9
tera	T	10^{12}
peta	P	10^{15}
exa	E	10^{18}
zeta	Z	10^{21}
yota	Y	10^{24}

SUBMULTIPLES USED IN THE SI

Prefix	Abbreviation	Scientific Notation
deci	d	10^{-1}
centi	c	10^{-2}
milli	m	10^{-3}
micro	μ	10^{-6}
nano	n	10^{-9}
pico	p	10^{-12}
femto	f	10^{-15}
atto	a	10^{-18}
zepto	z	10^{-21}
yocto	y	10^{-24}

APPENDIX B: SOME BASIC TERMS

Austenite is a dense face-centered cubic structure of iron that forms when **ferrite** is cooled to 1400 °C.

Basic oxygen furnace is used to convert **pig iron** to **steel**: it is a massive pear-shaped vessel lined with refractory material and mounted on trunions so it can be tilted for charging with **pig iron** and **steel scrap** and for discharging of liquid **steel**. Water-cooled oxygen lances are inserted through a removable lid.

Bessemer converter is a massive egg-shaped vessel that provided the first practical means of producing inexpensive **steel** from **pig iron** through **decarburization**. The converter was lined with a refractory material, and after it received hot metal through its open top, the charge was blasted for 15–30 min with cold air led through **tuyères**. The resulting exothermic oxidation removed carbon and silicon from the hot metal.

Billets are round or square-profile (up to 25 cm) pieces of steel produced by **continuous casting** that are converted to bars and rods.

Blast Large volume of air (>5000 m^3/min) is generated by a turbo blower (cold blast) and then led to **hot stoves**, where its temperature is raised to as much as 1300 °C before it is forced through **tuyères** into a blast furnace to supply oxygen for carbon oxidation and heat for iron ore smelting.

Blast furnace is a tall, conical structure set on massive foundations. Its parts, proceeding upwards, are: circular hearth with water-cooled **tuyères** situated along its perimeter; bosh, a short, truncated, and slightly outward-sloping cone; belly, the widest section; stack (shaft), the longest and slightly narrowing section where descending ore, coke, and flux are heated by ascending hot, CO-rich gases that reduce the oxides; and finally the throat, surmounted by a top cone. **Blast** for the smelting is supplied from adjacent **hot stoves**. Modern furnaces can operate for about two decades after they are blown-in and before they are left to cool for relining. Molten metal **(pig iron)** is periodically released through tapholes and removed for **casting**, and **slag** is taken out through cinder notches.

Blooms are rolled steel pieces with rectangular profile blooms (typically 40 × 60 cm) that are turned into structural shapes (beams), rails, and pipes.

Carbon appears as an interstitial impurity in iron, forming solid solutions with the three microstructural phases of **iron—α-ferrite**, **δ-ferrite**, and **γ-ferrite**—and combining with iron to form **cementite**. In low concentrations the element hardens **steels**, while its higher concentrations in **cast iron** make the metal brittle. Its removal from **cast iron** is done through **decarbonization**.

Cast iron (pig iron) is the name given to a group of metals produced in **blast furnaces** containing at least 95% and up to more than 97% **iron**. All of them have high carbon content, at least 1.8% for white cast iron but mostly between 2.5% and 4% for grey cast iron and ductile cast iron. The third most abundant element in unalloyed cast irons is silicon (0.5–3%) and the three elements are forming ternary Fe–C–Si alloy. Pure **iron** melts at 1523 °C, while melting temperatures of different kinds of cast iron are between 1150 and 1200 °C. **Cast iron** usually contains traces of Mn (0.1–1%), S (0.01–0.25%), and P (0.011%). High **carbon** content makes cast iron brittle and inferior to other common alloys. Cast iron's tensile strength is just 150–400 MPa, much less than that of bronze or brass; its **impact strength** is also very low, as is its **ductility**, but it has good strength in compression.

Cementite is a very hard intermetallic compound (Fe_3C, ferric carbide) which makes **cast iron** hard and brittle.

Continuous casting eliminated energy-wasting batch production, with **steel** cast first into ingots before its further processing. The idea originated with Henry Bessemer in the 1860s, but it became a commercial reality only during the 1950s, and now virtually all modern steel is cast continuously to produce **slabs**, **blooms**, and **sheets**.

Decarburization (decarbonization) is a process removing carbon from **cast iron**. This was done traditionally by laborious **puddling**, since the 1860s it has been accomplished in **Bessemer converters**. **Open hearth furnaces** were dominant until after WW II, and now the process is done either in **basic oxygen furnaces** or in **electric arc furnaces**.

Direct reduction of iron produces the metal by reacting iron ore (usually pellets) with a reducing agent (most commonly gas) at temperatures below iron's melting point, yielding solid sponge or briquettes.

Ductility measures the degree to which a material can deform under tensile stress. Ductile (malleable) metals can be hammered out or pressed into sheets, drawn into wires, and molded into shapes. High-strength ductile sheet **steel** is the most important material for modern car bodies.

Electric arc furnace is a massive tiltable container lined with refractory material whose charge (mostly scrap metal) is heated by an arc between large carbon electrodes inserted through its top. The furnace produces molten steel in batches of tens to hundreds of tonnes after heats typically lasting 30–45 min.

Eutectic point indicates the chemical composition and temperature of the lowest melting point for a mixture of components. Eutectic points of iron–carbon mixtures can be found on liquid–solid phase diagrams (composition on the horizontal axis, temperature on the vertical axis).

Ferrite is a body-centric cubic structure of **iron** formed as the metal is cooled below 1538 °C. Further cooling produces a denser face-centered cubic structure of **austenite**.

Ferrous alloys are, in the broadest sense, any **iron**-based alloys including all varieties of **steel**. According to a narrower definition ferroalloys combine **iron** with relatively high shares of other elements, including aluminum, chromium, manganese, and silicon.

Flux is crushed limestone or dolomite (or their combination) charged into a **blast furnace** together with **iron ore** and **coke**. Its melting removes sulfur and other impurities from the molten metal, and it forms metallurgical **slag**.

Impact strength is a standard measure of toughness. The Charpy V-notch test is used to measure energy needed to break a small notched specimen (at a given temperature) by a single impact from a pendulum. The minimal impact energy of **steel** ranges between 27 and 40 J.

Iron is the most common heavy metal found in the Earth's crust and its fourth most abundant constituent, following O_2, Si, and Al. **Iron** is also present in the planet's core, and it reaches us in many meteorites; it is the 26th element of the periodical table, with specific density of 7.87 g/cm^3, three times heavier than aluminum but about 12% lighter than copper and 36% lighter than lead. **Iron**, **steels**, and **cast iron** are three types of **ferrous alloys**; they are defined by their **carbon** content and by the shares of microstructural forms of **iron**. There are three principal microstructural components: α-ferrite has a body-centric cubic atomic structure that is stable at ambient temperature and transforms into γ-ferrite at 912 °C; **δ-ferrite** is structurally identical with α-ferrite, but it is stable only at temperatures above 1394 °C and it melts at 1538 °C; and the γ-ferrite (austenite) phase of carbon steel (up to 2.14% C) can exist only at high temperatures (above 727 °C), and it has a face-centered cubic atomic structure. **Iron** as α-ferrite contains only a negligible trace of carbon (0.008%) at ambient temperature, **steels** contain 0.008–2.14% C (but usually <1%) by mass and the rest is α-ferrite and **cementite**, and **cast irons** have 2.14–6.7% C (but usually <4.5%) by mass, with the rest being **cementite**.

Martensite is a body-centric tetragonal structure that arises from rapid cooling of **austenite**. This supersaturated solution of carbon in iron imparts strength and toughness to **steel**.

Open hearth furnace was a large, shallow basin lined with refractories and covered with a low arched roof. The furnace received molten pig iron, scrap, and fluxing materials and removed excessive carbon during a lengthy period of heating by a gaseous fuel.

Pellets of iron are made from ground iron fines mixed with fluxing materials and binders; coke or coal can also be added to aid the firing process. Pellets with diameters of 9–16 mm are processed in induration machines under temperatures of up to 1350 °C.

Pig iron is the term used interchangeably with **cast iron**: as the International Iron Metallics Association explains, the hot metal was traditionally cast into multiple sand molds branching from a central channel, a configuration resembling a litter of piglets suckling a sow, and after cooling these small ingots (the pigs) were broken off, cooled, and used for further processing (IIMA, 2014).

Plates Flat hot-rolled pieces (2–20 mm thick, up to 1.86 m wide) used to make structural steel products, ship hulls, boilers, pressure vessels, pipes, and various heavy metal structures, including offshore drilling platforms.

Puddling is a labor-intensive and very strenuous process to make **wrought iron** through **decarburization** of **cast (pig) iron** by stirring the hot iron bath and forming heavy balls of malleable metal.

Reduction Removal of oxygen in a chemical reaction. Smelting of ores and production of iron in **blast furnaces** is the world's most common chemical reduction. The principal process of **direct reduction** is $Fe_2O_3 + 3C \rightarrow 2Fe + 3CO$, while indirect reduction taking place in a **blast furnace** proceeds in a series of reactions:

$$3Fe_2O_3 + CO = CO_2 + 2Fe_3O_4$$
$$Fe_3O_4 + CO = CO_2 + 3FeO$$
$$O + CO = CO_2 + Fe$$

Reinforcing bars Hot-rolled **steel** pieces cut into various lengths to be used in strengthening concrete in buildings, highways, runways, dams, and bridges.

Reverberatory furnace is designed in such a way that the fuel and the molten metal are separated: usual arrangement is with heat reverberating (reflected from) a low sloping furnace roof and sides.

Rolling is a process that has been used for more than a century to turn **steel** into semifinished shapes, and now also into finished products. Originally it began with cast steel **ingots**; now it is a part of **continuous casting**. Three basic kinds of semifinished products are **slabs**, **billets**, and **blooms**, but continuous casting can also produce thins slabs and thin strips. Rolling can be done hot or cold. Hot-rolled coils are strips 2–7 mm thick and between 0.6 and 1.2 m wide that are used (with or without further processing) in construction, transportation, and pipelines.

Scrap is any metallic waste material suitable for recycling (after sorting and shredding). **Steel** scrap is the principal feed for electric arc furnaces.

Sections Hot-rolled pieces including beams and sheet piles used in buildings and in industrial and highway construction.

Slabs are wide and thick pieces of steel with rectangular cross-sections, now produced by **continuous casting** and then converted by further rolling into thin slabs, sheets, plates, pipes, and tubes.

Slag is a by-product of **pig iron** production, formed by melting **iron ore** (or its **pellets**), coke, and a **flux** (limestone or dolomite) in a **blast furnace** as lime in the flux combines with the aluminates and silicates in the ore and with **coke** ash. Molten slag can be cooled in different ways to produce several kinds of solid material used in cement production, in construction, and as a fertilizer.

Steel is, much like coal or crude oil, a singular describing a variety of materials of different compositions and different physical properties. The plural is thus a more accurate choice: **steels** are **ferrous alloys** characterized above all by their restricted **carbon** content, between 0.08% and 2.14% by weight, and high to very high **tensile strength** and high **yield strength** and **impact strength**. The addition of other elements—most commonly Al, Cr, Co, Mn, Mo, Ni, Ti, V, and W, in total amounts ranging from less than 2% to more than 10% of the mass—produces steel alloys with a variety of other desirable properties. Low-carbon sheet steel goes into car bodies, while highly tensile and hardened (with Mn, Mo, and Ni) steels go into axles, shafts, connecting rods, and gear. Stainless steels are indispensable for medical devices and household appliances as well as for chemical and food-processing equipment, and tool steels are made into thousands of devices ranging from chisels and gauges to extrusion dies and sheer blades.

Tensile strength measures the force used to pull a material to the point it can withstand without permanent deformation (yield strength) or to its breaking point (ultimate strength). The two points for structural steel are 250 and 400 MPa; for high-tensile steels they are up to 1.65 and 2.2 GPa.

Tuyère is a water-cooled nozzle (pipe) placed at the top of a **blast furnace** hearth and used to force blast from **hot stoves** into the furnace. Tuyères were also used to blow air into **Bessemer converters**, and they deliver oxygen through the bottom of **basic oxygen furnaces**.

Wrought iron contains only traces of **carbon** (0.04–0.08%), but it has inclusions of **slag** that produce its fibrous or mottled look. The traditional process of producing **wrought iron** was by **decarburization** of **pig iron**. **Wrought iron** forged (hot-worked) at different temperatures is softer and more malleable than modern steels and also more rust-resistant. In the preindustrial world, the metal was used widely to make items ranging from nails and chains to horseshoes and bolts, and before the adoption of the **Bessemer converter** to produce cheap **steel**, it was used to make rails for early railroads (as well as the Eiffel Tower).

APPENDIX C: GLOBAL AND NATIONAL PRODUCTION OF PIG IRON AND STEEL, 1800–2015

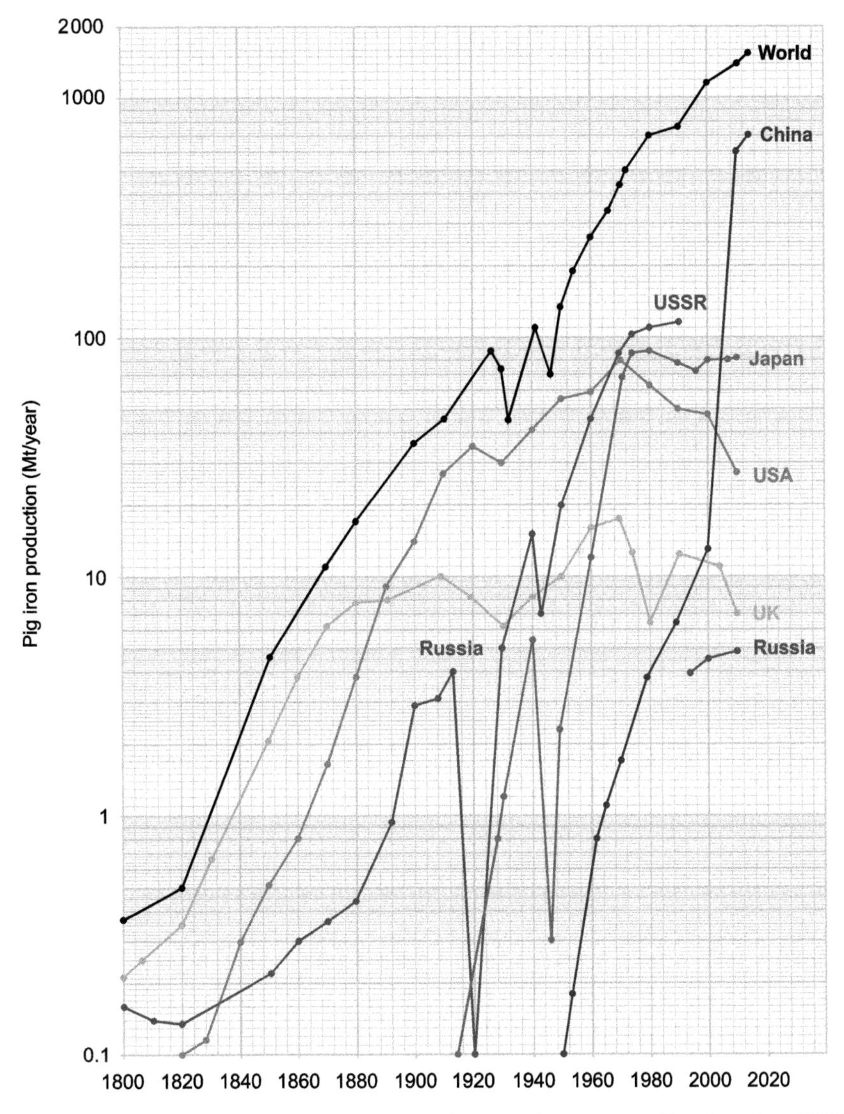

Figure C.1 Pig iron production. *Plotted from data in Palgrave Macmillan (2014) and WSA (2014).*

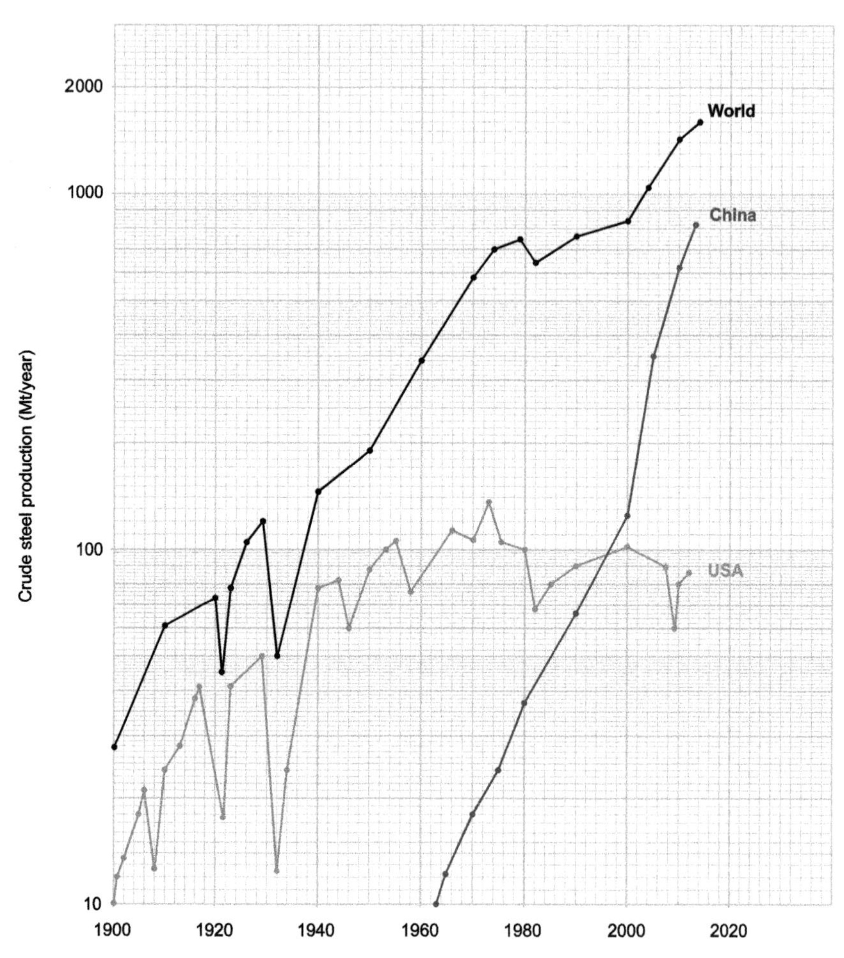

Figure C.2 Crude steel production. *Plotted from data in Palgrave Macmillan (2014) and WSA (2014).*

APPENDIX D: PRODUCTION OF CRUDE STEEL, 1900–2014 (ALL FIGURES IN Mt/YEAR)

	World	USA	Germany	UK	Russia	Japan	China
1900	28	9.2	6.5	5.0	2.2		
1910	60	23.7	13.1	6.5	3.3		
1920	73	37.8	9.3	9.2	0.2	0.8	
1930	95	36.4	12.5	7.4	5.8	2.3	
1940	141	78.0	21.5	13.2	18.3	6.9	
1950	192	87.8	13.1	16.6	27.3	4.8	0.1
1960	346	90.1	37.9	24.7	65.3	22.1	0.5
1970	595	119.0	50.0	28.3	115.9	93.3	17.8
1980	716	101.0	51.1	11.3	147.9	111.4	37.1
1990	771	89.7	38.4	17.8	154.4	110.3	66.4
2000	850	102.0	30.8	15.2	59.1	106.4	128.5
2010	1430	80.5	43.8	9.7	66.9	109.6	638.7
2013	1642	86.9	42.6	11.9	68.9	110.6	822.7
2014	1662	88.3	42.9	12.1	70.7	110.7	822.7

Source: Palgrave Macmillan (2013); WSA (1982, 2015).

REFERENCES

Adams, W., & Dirlam, J. B. (1966). Big steel, invention, and innovation. *The Quarterly Journal of Economics, 80*, 167–189.

Adriaanse, A., Bringezu, S., & Moriguchi, Y., et al. (1997). *Resource flows: The material basis of industrial economies*. Washington, DC: World Resources Institute.

Agricola, G. (1556). *De re metallica*. Basel: Froben. English translation by H. C. Hoover and L. H. Hoover, London: The Mining Magazine (1912).

AIG. (2015). Infographic: An American bridge's foreign components. *Quartz*, May 11, 2015, <http://qz.com/384285/infographic-an-american-bridges-foreign-components/>.

AISI (American Iron and Steel Institute). (2005). Saving One Barrel of Oil per Ton (SOBOT). <http://www.steel.org/~/media/Files/AISI/Public%20Policy/saving_one_barrel_oil_per_ton.pdf>.

AISI. (2009). Steel takes LEED© with recycled content. <http://www.recycle-steel.org/~/media/Files/SRI/Media%20Center/LEED_Sept2011.ashx>.

AISI. (2010). *Technology roadmap research program for the steel industry*. <http://www.steel.org/~/media/Files/AISI/Making%20Steel/TechReportResearchProgramFINAL.pdf>.

AISI. (2014). *Profile 2014*. <http://www.steel.org/~/media/Files/AISI/Making%20Steel/TechReportResearchProgramFINAL.pdf>.

AISI. (2015). 2015 steel industry profile. <https://www.steel.org/About%20AISI/Industry%20Profile.aspx>.

Allen, R. C. (1979). International competition in iron and steel, 1850–1913. *The Journal of Economic History, 39*, 911–937.

Allen, R. C. (1981). Entrepreneurship and technical progress in the northeast coast pig iron industry: 1850–1913. *Research in Economic History, 6*, 35–71.

Almond, J. K. (1981). A century of basic steel: Cleveland's place in successful removal of phosphorus from liquid iron in 1879, and development of basic converting in ensuing 100 years. *Ironmaking and Steelmaking, 8*, 1–10.

Alstom. (2013). Alstom boilers. <http://www.alstom.com/Global/Germany/Resources/Documents%20-%20Brochures/Brochure_Power_Boilers_forsteampowerplants-boilers.pdf>.

Anameric, B., & Kawatra, S. K. (2015). Direct iron smelting reduction processes. *Mineral Processing and Extractive Metallurgy Review: An International Journal, 30*, 1–51.

Antal, M. J., & Grønli, M. (2003). The art, science and technology of charcoal production. *Industrial and Engineering Chemical Research, 42*, 1619–1640.

Apelt, B. (2001). *The corporation: A centennial biography of United States steel corporation, 1901–2001*. Pittsburgh, PA: Cathedral Publishing.

ArcelorMittal. (2014). ArcelorMittal signs a new contract with STX France shipyards for three new cruise ships. <http://flateurope.arcelormittal.com/dec/stxnewcontract>.

ArcelorMittal. (2015). ArcelorMittal. <http://corporate.arcelormittal.com/who-we-are/our-history>.

Arens, M., Worrell, E., & Schleich, J. (2012). Energy intensity development of the German iron and steel industry between 1991 and 2007. *Energy Policy, 45*, 786–797.

Arpi, G. (1953). The supply with charcoal of the Swedish iron industry from 1830 to 1950. *Geografiska Annaler, 35*, 11–27.

ASCE (American Society of Civil Engineers). (2015). *2013 Report Card for America's Infrastructure*. <http://www.infrastructurereportcard.org/grades/>.

Ashkenazi, D., Golan, O., & Tal, O. (2013). An archaeometallurgical study of 13th-century arrowheads and bolts from the crusader castle of Arsuf/Arsur. *Archaeometry, 55*, 235–257.

ASME (The American Society of Mechanical Engineers). (1985). *Oxygen process steel-making vessel*. New York, NY: ASME. <https://www.asme.org/getmedia/2561f8b5-a343-4d25-bef2-9459b23525e8/104_Basic_Oxygen_Steel_Making_Vessel_1955.aspx>.

Ausubel, J. H., & Waggoner, P. E. (2008). Dematerialization: Variety, caution, and persistence. *Proceedings of the National Academy of Sciences*, *105*, 12774–12779.

Avery, D. H., & Schmidt, P. R. (1979). A metallurgical study of the iron bloomery, particularly as practiced in Buhaya. *Journal of Metals*, *31*(9), 14–20.

Aylen, J. (2002). *The Continuing Story of Continuous Casting*. Keynote address to 4th European Continuous Casting Conference, Birmingham.

Babcock, F. L. (1931). *Spanning the Atlantic*. New York, NY: Alfred A. Knopf.

Babich, A., Senk, D., & Fernandez, M. (2010). Charcoal behavior by its injection into the modern blast furnace. *ISIJ International*, *50*, 81–88.

Bagsarian, T. (1998). Unveiling Project M. *New Steel*, 1–5. December 1998.

Bagsarian, T. (2000). Strip casting gets serious. *New Steel*, 1–7. December 2000.

Bagsarian, T. (2001). Blast furnace's next frontier: 20-year campaigns. *New Steel*, 1–6. July 2001.

Bailis, R., et al. (2013). Innovation in charcoal production: A comparative life-cycle assessment of two kiln technologies in Brazil. *Energy for Sustainable Development*, *17*, 189–200.

Baltic, S., & Hilliard, D. (2013). Steel vs aluminum in automotive lightweighting. *Insight Global Metals Trends*, 9–12. May 2013.

Barrena, M. I., Gómez de Salazar, J. M., & Soria, A. (2008). Roman iron axes manufacturing technology. *Nuclear Instruments and Methods in Physics Research B*, *266*, 955–960.

Bayley, J., Dungworth, D., & Paynter, S. (2001). *Archaeometallurgy*. London: English Heritage.

BEA (Bureau of Economic Analysis). (2015). Gross Domestic Product (GDP). <http://www.bea.gov/national/index.htm#gdp>.

Becher, B., & Becher, H. (1990). *Blast furnaces*. Cambridge, MA: MIT Press.

Becher, B., & Becher, H. (2002). *Hochöffen*. München: Schirmer/Mosel.

Beckett, C. A. (2012). Thomas Edison's beautiful failure. <http://christineadamsbeckett.com/2012/04/03/thomas-edisons-beautiful-failure/>.

Belford, P. (2010). Five centuries of iron working: Excavations at Wednesbury Forge. *Post-Medieval Archaeology*, *44/1*, 1–53.

Bell, L. (1884). *Principles of the manufacture of iron and steel*. London: George Routledge & Sons.

Berry, B., Ritt, A., & Greissel, M. (1999). A retrospective of twentieth-century steel. *Iron Age New Steel*, 1–14. November 1999.

Berry, R. S., & Fels, M. (1973). The energy cost of automobiles. *Bulletin of the Atomic Scientists* December 1973: 11–17, 58–60.

Bessemer, H. (1891). On the manufacture of continuous sheets of malleable iron and steel, direct from the fluid metal. *Journal of the Iron and Steel Institute*, *6*(10), 23–41.

Besta, P., et al. (2012). The effect of harmful elements in production of iron in relation to input and output material balance. *Metalurgija*, *51*, 325–328.

BIR (Bureau of International Recycling). (2014). *World steel recycling in figures 2009–2013*. Brussels: BIR. <http://www.bir.org/assets/Documents/publications/brochures/Ferrous-report-2014-LightWeb.pdf>

Birat, J.-P. (2010). ULCOS program: Status and progress. <http://www.eesc.europa.eu/resources/docs/estep_ulcos_nov_2010.pdf>.

Birch, A. (1967). *The economic history of the British iron and steel industry 1784–1879*. London: Cass.

Birch, A. (1968). *The economic history of the british iron and steel industry 1784–1879*. London: Cass.

Biringuccio, V. (1540). *De la pirotechnia*. English translation: *The Pirotechnia*, translated by C. S. Smith and M. T. Gnudi, New York, NY: Basic Books, 1959.

Biswas, A. K. (1994). Iron and steel in pre-modern India: A critical review. *Indian Journal of History of Science*, *29*, 579–610.

Blaenavon World Heritage Site. (2015). The Gilchrist–Thomas process. <http://www.visitblaenavon.co.uk/en/WorldHeritageSite/TheBlaenavonStory/BlaenavonandtheGilchrist-ThomasProcess.aspx>.

Boesenberg, J. S. (2006). Wrought iron from the *USS Monitor*: Mineralogy, petrology and metallography. *Archaeometry*, *48*, 613–631.

Bone, W. A. (1928). The centenary of James B. Neilson's invention of hot-blast in iron smelting. *Nature*, *122*, 317–319.

Börjesson, P., & Gustavsson, L. (2000). Greenhouse gas balances in building construction: Wood versus concrete from life-cycle and forest land-use perspective. *Energy Policy*, *28*, 575–588.

Boustead, I., & Hancock, G. F. (1979). *Handbook of industrial energy analysis*. Chichester: Ellis Horwood.

Bowman, J. (1989). *Andrew Carnegie: Steel tycoon*. Englewood Cliffs, NJ: Silver Burdett.

Boylan, M. (1975). *Economic effects of scale increases in the steel industry: The case of U.S. blast furnaces*. New York, NY: Praeger.

Boylston, H. M. (1936). *An introduction to the metallurgy of iron and steel*. New York, NY: John Wiley.

BP (British Petroleum), 2015. BP Statistical Review of World Energy. <http://www.bp.com/content/dam/bp/pdf/Energy-economics/statistical-review-2015/bp-statistical-review-of-world-energy-2015-full-report.pdf>.

Brazilian Mining Association, 2015. Annual Report. <http://www.ibram.org.br/sites/1400/1457/00000383.pdf>.

Brown, H. L., Hamel, B. B., & Hedman, B. A. (1996). *Energy analysis of 108 industrial products*. Lilburn, GA: Fairmont Press.

Brown, S., Schroeder, P., & Birdsey, R. (1997). Aboveground biomass distribution of US eastern hardwood forests and the use of large trees as an indicator of forest development. *Forest Ecology and Management*, *96*, 37–47.

Brunke, J.-C., & Blesl, M. (2014). A plant-specific bottom-up approach for assessing the cost-effective energy conservation potential and its ability to compensate rising energy-related costs in the German iron and steel industry. *Energy Policy*, *67*, 431–446.

Bryson, W. E. (2005). *Heat treatment, selection, and application of tool steels*. Cincinnati, OH: Hanser Gardner Publications.

Buchwald, V. F. (1998). Slag analysis as a method for the characterization and provenancing of ancient iron objects. *Materials Characterization*, *40*, 73–96.

Buckingham, D. A. (2006). Steel stocks in use in automobiles in the United States. <http://pubs.usgs.gov/fs/2005/3144/fs2005_3144.pdf>.

Bulasová, A. et al. (2014). Projekt kubismus. <http://www.sgjs.cz/assets/files/projekty/kubismus.pdf>.

Burchart-Korol, D. (2013). Life cycle assessment of steel production in Poland: A case study. *Journal of Cleaner Production*, *54*, 235–243.

Burchart-Korol, D., & Kruczek, M. (2015). Water scarcity assessment of steel production in national integrated steelmaking route. *Metalurgija*, *54*, 276–278.

Burgo, J. A. (1999). The manufacture of pig iron in blast furnace. In D. A. Wakelin (Ed.), *The making, shaping and treating of steel, ironmaking volume* (pp. 699–739). Pittsburgh, PA: The AISE Foundation.

Burton, R. F. (1884). *The book of the sword*. London: Chatto and Windus.

Burwell, C. C. (1990). High-temperature electroprocessing: Steel and glass. In S. H. Schurr et al., (Eds.), *Electricity in the American Economy* (pp. 109–129). New York: Greenwood Press.

Campbell, H. R. (1907). *The manufacture and properties of iron and steel*. New York, NY: Hill Publishing.

Cannon, W. F. (2011). The Lake Superior Iron Ranges: Geology and Mining. <http://mn.water.usgs.gov/projects/tesnar/2011/Presentations/CannonIron%20ranges.pdf>.

Carlton, D., & Perloff, J. (2005). *Modern industrial organisation.* Upper Saddle River, NJ: Pearson.
Carnegie Steel Company. (1912). *Carnegie steel company: General statistics and special treatise on homestead steel works.* Pittsburgh, PA: Carnegie Steel Company. <http://hdl.handle.net/10493/525>.
Caron, F. (2013). *Dynamics of innovation: The expansion of technology in modern times.* New York, NY: Berghahn.
CCTV. (2014). China steel price. October 17, 2014, <http://newscontent.cctv.com/NewJsp/news.jsp?fileId=264770>.
Chen, W., & Graedel, T. E. (2012). Anthropogenic cycles of the elements: A critical review. *Environmental Science & Technology, 46,* 8574–8586.
Chen, W., Yin, X., & Ma, D. (2014). A bottom-up analysis of China's iron and steel industrial energy consumption and CO_2 emissions. *Applied Energy, 136,* 1174–1183.
Chernykh, E. N. (2014). Metallurgical provinces of Eurasia in the early metal age: Problems of interrelation. *ISIJ International, 54,* 1002–1009.
CLIA (Cruise Line International Association). (2104). Cruise industry investment in ship innovations. <http://www.cruising.org/vacation/news/press_releases/2014/01/state-cruise-industry-2014-global-growth-passenger-numbers-and-product-o>.
CMI (Can Manufacturers Institute). (2013). Food cans overview. <http://www.cancentral.com/food-cans/overview>.
CNI (National Confederation of Industry). 2012. *Steel industry in Brazil.* <http://ibnbio.org/wp-content/uploads/2012/10/ING_CNI_PARTE2_RIO20.pdf>.
Cobb, H. M. (2010). *The history of stainless steel.* Materials Park, OH: ASM International.
Cossons, N., & Trinder, B. (1979). *The iron bridge: Symbol of the industrial revolution.* Bradford on Avon: Moonraker Press.
Cowper, E. A. (1866). On the effect of blowing blast furnaces with blast of very high temperatures. In: *Report of the 35th meeting of the British Association for the advancement of science.* London: John Murray, p. 177.
Craddock, P. T. (1995). *Early metal mining and production.* Edinburgh: Edinburgh University Press.
Credit Suisse. (2012). Have we reached "Peak Steel" demand in China? We think not. <https://doc.research-and-analytics.csfb.com/docView?language=ENG&format=PDF&document_id=804893350&source_id=em&serialid=ljR8%2FPmJnc6nRhduTod7Cl3gVu2YcKPXFh6f2SZ0pvc%3D>.
Crossley, D. W. (Ed.), (1981). *Medieval industry.* London: Council for British Archaeology.
Cullen, J. M., et al. (2012). Mapping the global flow of steel: From steelmaking to end-use goods. *Environmental Science & Technology, 46,* 13048–13055.
Dahlström, K., et al. (2004). *Iron, steel and aluminium in the UK: Material flows and their economic dimensions.* Guildford: University of Surrey.
Dai, Y., & Gutierrez, T. (2014). Smog choking Chinese steel. *Insight,* 4–7. May 2014.
Daimler. (2015). The birth of the automobile. The 35-hp Mercedes, the first modern automobile (1900–1901). <http://www.daimler.com/dccom/0-5-1322446-1-1323365-1-0-0-1322455-0-0-135-0-0-0-0-0-0-0-0.html>.
Danieli Corus. (2014). Pulverized coal injection. <http://www.danieli-corus.com/media/PCI.pdf>.
Danloy, G. et al. (2008). Heat and mass balance in the ULCO blast furnace. *Proceedings of the 4th ULCOS Seminar, October 1–2, 2008.* <http://www.ulcos.org/en/docs/seminars/Ref11%20-%20SP10_Danloy1_Essen_New.pdf>.
Dartnell, J. (1978). Coke in the blast furnace. *Ironmaking and Steelmaking, 1978*(1), 18–21.
Davis, J. J. (1922). *The iron puddler, my life in the rolling mills and what came of it.* New York, NY: Grosset and Dunlap.

De Beer, J., Worrell, E., & Blok, K. (1998). Future technologies for energy-efficient iron and steel making. *Annual Review of Energy and the Environment, 23*, 123–205.

DECC (Department of Energy & Climate Change). (2014). Historical coal data: Coal production, 1853 to 2013. <https://www.gov.uk/government/statistical-data-sets/historical-coal-data-coal-production-availability-and-consumption-1853-to-2011>.

Derui, T., & Haiping, L. (2011). The ancient Chinese casting techniques. *Foundry World, 8*, 127–133. <http://www.foundryworld.com/uploadfile/201131449329893.pdf>.

De Ryck, I., Adriaens, A., & Adams, F. (2005). An overview of Mesopotamian bronze metallurgy during the 3rd millennium BC. *Journal of Cultural Heritage, 6*, 261–268.

Diderot, D., & D'Alembert, J. L. R. (1751–1777). *L'Encyclopedie ou dictionnaire raisonne des sciences des arts et des metiers*. Paris: Avec Approbation and Privilege du Roy.

Díez, M. A., Alvarez, R., & Barriocanal, C. (2002). Coal for metallurgical coke production: predictions of coke quality and future requirements for cokemaking. *International Journal of Coal Geology, 50*, 389–412.

Dogs of the Dow. (2015). <http://www.dogsofthedow.com/djdelete.htm>.

Domergue, C. (1993). *Un centre sidérurgique romain de la Montagne noire. Le Domaine des forges (Les Martys, Aude)*. Paris: Editions du CNRS.

Drougas, A. (2009). *Investigation of the use of iron in construction from antiquity to the technical revolution*. Barcelona: Universitat Politecnica de Catalunya.

Dumpleton, B., & Miller, M. (1974). *Brunel's three ships*. Melksham: Colin Venton.

Dunayevskaya, R. (1942). An analysis of Russian economy. *The New International, 8*(11), 327–332.

Egenhofer, C., et al. (2013). *The steel industry in the European Union: Composition and drivers of energy prices and costs*. Brussels: Center for European Policy Studies. <http://aei.pitt.edu/46349/1/Steel_Report_(1).pdf>

Egerton, W. (1896). *Indian and oriental armour*. London: W.H. Allen.

Elbaum, B. (2007). How godzilla ate Pittsburgh: The long rise of the Japanese iron and steel industry, 1900–1973. *Social Science Japan Journal, 10*, 243–264.

Emerick, H. B. (1954). European oxygen steelmaking is of far-reaching significance. *Journal of Metals, 6*, 803–805.

Emi, T. (2015). Steelmaking technology for the last 100 years: Toward highly efficient mass production systems for high quality steels. *ISIJ International, 55*, 36–66.

Emmerich, F. G., & Luengo, C. A. (1996). Babassu charcoal: A sulfurless renewable thermo-reduction feedstock for steelmaking. *Biomass and Bioenergy, 10*, 41–44.

Evans, C., Jackson, O., & Rydén, G. (2002). Baltic iron and the British iron industry in the eighteenth century. *Economic History Review, 55*, 642–665.

Evelyn, J. (1664). *Sylva or a discourse of forest-trees and the propagation of timber*. London: John Martyn.

FAO. (1983). Use of charcoal in blast furnace operations. <http://www.fao.org/docrep/03500e/03500e07.htm>.

FAO. (2014). *Forest products 2008–2012*. Rome: FAO.

Felkins, K., Leighly, H. P., & Jankovic, A. (1998). The royal mail ship Titanic: Did a metallurgical failure cause a night to remember? *JOM, 50*, 12–18.

Fell, A. (1908). *The early iron industry of furness and district*. London: Frank Cass.

Fenske, G. (2008). *The skyscraper and the city: The Woolworth building and the making of modern New York*. Chicago: University of Chicago Press.

Ferreira, O. C. (2000). The future of charcoal in metallurgy. *Energy & Economy, 21*, 1–5.

Fenton, M. D. (2014). Iron and steel scrap in December 2014. <http://minerals.usgs.gov/minerals/pubs/commodity//iron_&_steel_scrap/mis-201412-fescr.pdf>.

Feuerbach, A. (2006). Crucible damascus steel: A fascination for almost 2,000 years. *Journal of Metals May, 2006*, 48–50.

Feuerstein, G. (1998). Brooklyn Bridge facts, history, and information. <http://web.archive.org/web/20100208205253/http://www.endex.com/gf/buildings/bbridge/bbridgefacts.htm>.

Field, F. F., et al. (1994). *Automobile recycling policy: Background materials*. Davos: International Motor Vehicle Program.

Figiel, L. S. (1991). *On damascus steel*. Atlantis, FL: Atlantis Art Press.

Fischedick, M., et al. (2014). Techno-economic evaluation of innovative steel production technologies. *Journal of Cleaner Production, 84*, 563–580.

Flandrin, J. (1989). Distinction through taste. In P. Ariès & G. Duby (Eds.), *A history of private life III passions of the renaissance* (pp. 265–307). Cambridge, MA: Harvard University Press.

Flemings, M. C., & Ragone, D. V. (2009). Puddling: A new look at an old process. *ISIJ International, 49*, 1960–1966.

Ford. (1909). *Watch the Ford Go By*. <http://www.thehenryford.org/exhibits/showroom/1908/lit.html>.

Forth Bridges. (2013). Forth bridges. <http://www.forth-bridges.co.uk/>.

Fruehan, R. J., et al. (2000). *Theoretical minimum energies to produce steel for selected conditions*. Columbia, MD: Energetics. <http://energy.gov/sites/prod/files/2013/11/f4/theoretical_minimum_energies.pdf>

Fruehan, R. J. (2005). New steelmaking processes: Drivers, requirements and potential impact. *Ironmaking and Steelmaking, 32*, 3–8.

Fry, H. (1896). *The history of North Atlantic steam navigation: With some account of early ships and shipowners*. London: Sampson Low, Marston and Company.

Fujitsuka, N. et al. (2013). Dynamic modeling of world steel cycle toward 2050. <http://conferences.chalmers.se/index.php/LCM/LCM2013/paper/viewFile/734/333>.

Galán, J., et al. (2012). Advanced high strength steels for automotive industry. *Revista de Metalurgia, 48*, 118–131.

Gale, J., & Freund, P. (2001). *Greenhouse gas abatement in energy intensive industries*. Paris: IEA.

Gao, C., et al. (2011). Optimization and evaluation of steel industry's water-use system. *Journal of Cleaner Production, 19*, 564–569.

Gatrell, P., & Harrison, M. (1993). The Russian and Soviet economy in two world wars. *Economic History Review, 46*, 425–452.

Geerdes, M., Toxopeus, H., & van der Vliet, C. (2009). *Modern blast furnace ironmaking*. Amsterdam: IOS Press.

Gervásio, H., et al. (2014). Comparative life cycle assessment of tubular wind towers and foundations—Part 2: Life cycle analysis. *Engineering Structures, 74*, 292–299.

Ghenda, J. T. (2014). Review of the EU ETS Post-2020: Impact assessment on the EU steel sector. <http://ec.europa.eu/transparency/regexpert/index.cfm?do=groupDetail.groupDetailDoc&id=14819&no=2>.

Ginley, D. S., & Cahen, D. (Eds.), (2012). *Fundamentals of materials for energy and environmental sustainability*. Cambridge: Cambridge University Press.

Godfrey, E., & van Nie, M. (2004). A Germanic ultrahigh carbon steel punch of the Late Roman-Iron Age. *Journal of Archaeological Science, 31*, 1117–1125.

Gold, B., et al. (1984). *Technological progress and industrial leadership: The growth of the U.S. steel industry, 1900–1970*. Lexington, MA: D.C. Heath and Company.

Goldsworthy, T., & Gull S. (2002). HIsmelt—The new technology for iron production. <http://www.riotinto.com/documents/_Iron%20Ore/HIsmelt_0402_the_new_technology_for_iron_production.pdf>.

González, R. F., et al. (2004). Stone decay in 18th century monuments due to iron corrosion. The Royal Palace, Madrid (Spain). *Building and Environment, 39*, 357–364.

Government of India. (2012). National Steel Policy 2012 (Draft). <http://steel.gov.in/06112012%20National%20Steel%20Policy%20Draft.pdf>.

Grazzi, F., et al. (2011). Ancient and historic steel in Japan, India and Europe, a non-invasive comparative study using thermal neutron diffraction. *Analytical and Bioanalytical Chemistry, 400*, 1493–1500.

Greenpeace. (2013). *Driving destruction in the Amazon: How steel production is throwing the forest into the furnace*. Amsterdam: Greenpeace International. <http://www.greenpeace.org/international/Global/international/publications/forests/2012/Amazon/423-Driving-Destruction-in-the-Amazon.pdf>.

Greenwood, W. H. (1907). *Iron*. London: Cassell and Company.

Greissel, M. (2000). The power of oxygen. *New Steel, 16*(4), 24–30.

Guglielmini, A., & Degel, R. (2007). *Direct ironmaking via rotary hearth furnace and new smelting technology*. Brussels: European Commission.

Guo, Z., & Xu, Z. (2010). Current situation of energy consumption and measures taken for energy saving in the iron and steel industry in China. *Energy, 35*, 4356–4360.

Haaland, R., & Shinnie, P. (Eds.), (1985). *African iron working—Ancient and traditional*. Oslo: Norwegian University Press.

Haapakangas, J., et al. (2011). A method for evaluating coke hot strength. *Fuel, 90*, 384–388.

Haga, T. (2004). *The latest trend of ironmaking technology in Japan: Relining of Oita No. 2 BF (The Largest BF in the world)*. Tokyo: Nippon Steel Corporation.

Halder, S. (2011). An experimental perspective on Praxair's hot oxygen technology to enhance pulverized solid fuel combustion for ironmaking blast furnaces. <http://www.praxair.com/~/media/North%20America/US/Documents/Reports%20Papers%20Case%20Studies%20and%20Presentations/Industries/Metal%20Production/Paper%202011%20AISTech%20Hot%20Oxygen%20for%20Coal%20Combustion%20Halder.pdf>.

Hall, C. G. L. (1997). *Steel phoenix: The fall and rise of the U.S. steel industry*. New York, NY: St. Martin's Press.

Haller, W. (2005). Industrial restructuring and urban change in the Pittsburgh region: developmental, ecological, and socioeconomic trade-offs. *Ecology and Society, 10*(1), 13. [online] URL: <http://www.ecologyandsociety.org/vol10/iss1/art13/>.

Hammersley, G. (1973). The charcoal iron industry and its fuel, 1540–1750. *Economic History Review, 24*, 593–613.

Hammervold, J., Reenaas, M., & Brattebø, H. (2013). Environmental life cycle assessment of bridges. *Journal of Bridge Engineering, 18*, 153–161.

Harada, T., & Tanaka, H. (2011). Future steelmaking model by direct reduction techniques. *ISIJ International, 51*, 1301–1307.

Harding, A. (2014). Rio Tinto: generating significant business value. <http://www.riotinto.com/documents/140311_Andrew_Harding_Presentation_slides_AJM_Global_Iron_Ore_and_Steel_Forecast_Conference(1).pdf>.

Harris, J. R. (1988). *The British iron industry 1700–1850*. London: Macmillan.

Hartman, J. M. (1980). *Regenerative stoves—A sketch of their history and notes on their use*. Englewood, CO: American Institute of Mining Engineers.

Hasanbeigi, A., et al. (2011). *A comparison of iron and steel production energy use and energy intensity in China and the US*. Berkeley, CA: Ernest Orlando Lawrence Berkeley National Laboratory. <https://china.lbl.gov/sites/all/files/lbl-4836e-us-china-steeljune-2011.pdf>.

Hasanbeigi, A., et al. (2014). Comparison of iron and steel production energy use and energy intensity on China and the US. *Journal of Cleaner Production, 65*, 108–119.

Hasanbeigi, A., Price, L., & Arens M. (2013). Emerging energy-efficiency and carbon dioxide emissions-reduction technologies for the iron and steel industry. Ernest Orlando Lawrence Berkeley National Laboratory, <https://china.lbl.gov/sites/all/files/6106e-steel-tech.pdf>.

Hatayama, H., et al. (2010). Outlook of the world steel cycle based on the stock and flow dynamics. *Environmental Science & Technology, 44*, 6457–6463.

Hattori., R., et al. (2013). Estimation of in-use steel stock for civil engineering and buildings using nighttime light images. *Resources, Conservation and Recycling, 31*, 58–68.

Heal, D. W. (1975). Modern perspectives on the history of fuel economy in the iron and steel industry. *Ironmaking and Steelmaking, 1975*(4), 222–227.

Herrigel, G. (2010). *Manufacturing possibilities: Creative action and industrial recomposition in the United States, Germany, and Japan*. New York, NY: Oxford University Press.

Hess, G. W. (1989). Is the blast furnace in its twilight? *Iron Age, 5*(11), 16–26.

Hessen, R. (1975). *Steel titan: The life of Charles M. Schwab*. New York, NY: Oxford University Press.

Heuss, R., et al. (2012). *Lighweight, heavy impact*. <www.mckinsey.com/~/media/.../Lightweight_heavy_impact.ash>.

Hidalgo, I. (2003). *Energy consumption and CO_2 emissions from the world iron and steel industry* (<http://ftp.jrc.es/EURdoc/eur20686en.pdf>). Brussels: EU. <http://ftp.jrc.es/EURdoc/eur20686en.pdf>.

Hitachi Metals. (2014). Tale of the Tatara. <http://www.hitachi-metals.co.jp/e/tatara/index.htm>.

Hoffmann, H. (1953). *Die chemische Veredlung der Steinkohle durch Verkokung*. <http://epic.awi.de/23532/1/Hof1953a.pdf>.

Hoffmann, O. (2012). Steel lightweight materials and design for environmental friendly mobility. <http://www.industrialtechnologies2012.eu/sites/default/files/presentations_session/03_Oliver_Hoffmann.pdf>.

Hogan, W. T. (1971). In *Economic history of the iron and steel industry in the United States* (vol. 5). Lexington, MA: Lexington Books.

Höganäs. 2015. Höganäs. <https://www.hoganas.com/>.

Holl, A. F. C. (2009). Early West African metallurgies: New data and old orthodoxy. *Journal of World Prehistory, 22*, 425–438.

Horie, S., et al. (2011). Comparison of water footprint for industrial products in Japan, China and USA. In M. Finkbeiner (Ed.), *Towards life cycle sustainability management* (pp. 155–160). Berlin: Springer Verlag.

HSBEC (Honshu-Shikoku Bridge Expressway Company). (2015). Introduction of Akashi-Kaikyo Bridge. <http://www.jb-honshi.co.jp/english/bridgeworld/bridge.html>.

Hsu, F., et al. (2011). Estimation of steel stock in building and civil construction by satellite images. *ISIJ International, 51*, 313–319.

Hsu, F., Elvidge, C. D., & Masuno, Y. (2015). Exploring and estimating in-use steel stocks in civil engineering and buildings from night-time lights. *International Journal of Remote Sensing, 34*, 490–504.

Hu, M., et al. (2010). Iron and steel in Chinese residential buildings: A dynamic analysis. *Resources, Conservation and Recycling, 54*, 591–600.

Hua, J. (1983). The mass production of iron castings in ancient China. *Scientific American, 248*, 120–128.

Huang, N. (1958). *China will overtake Britain*. Beijing: Foreign Languages Press.

Hudson, H. D. (1986). *The rise of the Demidov family and the Russian iron industry in the eighteenth century*. Newtonville, MA: Oriental Research Partners.

Hyde, C. K. (1977). *Technological change and the British iron industry 1700–1870*. Princeton, NJ: Princeton University Press.

IEA. (2010). Iron and steel. *IEA ETSAP Technology Brief 102* <http://www.etsap.org/E-techDS/PDF/I02-Iron&Steel-GS-AD-gct.pdf>.

IEA. (2012). *CO_2 abatement in the iron and steel industry*. London: IEA. <http://www.iea-coal.org.uk/documents/82861/8363/CO2-abatement-in-the-iron-and-steel-industry,-CCC/193>

IEA (International Energy Agency). (2008). Energy technology perspectives. <http://www.iea.org/media/etp/etp2008.pdf>.

IETD (Industrial Efficiency Technology Database). (2015). Blast furnace system. <http://ietd.iipnetwork.org/content/blast-furnace-system>.

IFIAS (International Federation of Institutes for Advanced Study). (1974). *Energy analysis workshop on methodology and conventions*. Stockholm: IFIAS.

Iida, K. (1980). *Origin and development of iron and steel technology in Japan*. Tokyo: The United Nations University.

IIMA (International Iron Metallics Association). (2014). Pig iron. <http://metallics.org.uk/pigiron/>.

IMF (International Monetary Fund). (2015). Principal global indicators. <http://www.principalglobalindicators.org/Pages/Default.aspx>.

IMO (International Maritime Organization). 2015. The Hong Kong International Convention for the Safe and Environmentally Sound Recycling of Ships. <http://www.imo.org/About/Conventions/ListOfConventions/Pages/The-Hong-Kong-International-Convention-for-the-Safe-and-Environmentally-Sound-Recycling-of-Ships.aspx>.

Ingham, J. N. (1978). *The iron barons: A social analysis of an American urban elite, 1874–1965*. Westport, CT: Greenwood Press.

INMETCO. (2015). INMETCO: Recycling to sustain our natural resources. <http://www.inmetco.com/>.

IPCC. (2006). *2006 Guidelines for national greenhouse gas inventories* (<http://www.ipcc-nggip.iges.or.jp/public/2006gl/>). Geneva: IPCC. <http://www.ipcc-nggip.iges.or.jp/public/2006gl/>

IPCC. (2007). *Contribution of working group I to the fourth assessment report of the intergovernmental panel on climate change*. Geneva: IPCC. <https://www.ipcc.ch/publications_and_data/ar4/wg1/en/contents.html>

ISO (International Organization for Standardization). (2006). *The new international standards for life cycle assessment*. Geneva: ISO.

ISSF (International Stainless Steel Forum). (2015). Stainless steel bars & wires in electronics. <http://www.worldstainless.org/Files/issf/Animations/IT/flash.html>.

Iwasaki, M., & Matsuo, M. (2012). Change and development of steelmaking technology. *Nippon Steel Technological Report*, *101*, 89–94. <http://www.nssmc.com/en/tech/report/nsc/pdf/NSTR101-09_tech_review-1-3.pdf>.

Jamasmie, C. (2014). Australia, Brazil to control 90% of global iron trade by 2020. <http://www.mining.com/australia-brazil-to-control-90-of-global-iron-ore-trade-by-2020-56972/>.

Jeans, W. (1884). *The creators of the age of steel*. London: Chapman and Hall.

Jensen, A. A., et al. (1998). *Life cycle assessment (LCA): A guide to approaches, experiences and information sources*. Copenhagen: European Environment Agency.

Jernkontoret. (2014). A glance at the Swedish steel industry. <http://www.met.kth.se/asialink/Curriculum/Royal%20Institute%20of%20Tech-KTH/Swedish%20steel%20industry%20in%20english-Birgita.pdf>.

JISF (The Japan Iron and Steel federation). (2015). Japanese steel production. <http://www.jisf.or.jp/en/statistics/production/index.html>.

Jody, B. J., et al. (2009). *Recycling end-of-life vehicles of the future*. <http://www.ipd.anl.gov/anlpubs/2010/01/65969.pdf>.

Johannsen, O. (1953). *Geschichte des Eisens*. Dusseldorf: Verlag Stahleisen.

Johnson, J., et al. (2008). The energy benefit of stainless steel recycling. *Energy Policy*, *36*, 181–192.

Jones, A. (2014). International market for metallurgical coke. Steelhome Annual Conference, Shanghai, April 2014.

Jones, J. A. T. (2003). *Electric arc furnace steelmaking*. Pittsburgh, PA: AISI.

Jones, J. A. T., Bowman, B., & Lefrank, P. A. (1998). Electric furnace steelmaking. In R. J. Fruehan (Ed.), *The making, shaping and treating of steel* (pp. 525–660). Pittsburgh, PA: The AISE Steel Foundation.

JRC (Joint Research Center). (2011). *2011 Technology map of the European Strategic Energy Technology Plan. Brussels: European Commission.* <https://setis.ec.europa.eu/system/files/Technology_Map_2011.pdf>.

Juleff, G. (1996). An ancient wind-powered iron smelting technology in Sri Lanka. *Nature, 379*, 60–63.

Juleff, G. (2009). Technology and evolution: A root and branch view of Asian iron from first-millennium BC Sri Lanka to Japanese steel. *World Archaeology, 41*, 557–577.

Kakela, P. J. (1981). Iron ore: From depletion to abundance. *Science, 212*, 132–136.

Kanno, R., et al. (2012). Steels, steel products and steel structures sustaining growth of society (infrastructure field). *Nippon Steel Technical Report, 101*, 57–67.

Kato, M., et al. (2005). World at work: Charcoal producing industries in northeastern Brazil. *Occupational and Environmental Medicine, 62*, 128–132.

Kato, K., et al. (2006). Nippon Steel Corporation developed a new technology for waste plastic recycling process using coke ovens. *Nippon Steel Technical Report, 94*, 75–79.

Kawahara, K., et al. (2012). Estimation of world steel demand and in-use stock for ships. *Tetsu to Hagane, 98*, 27–34.

Kawaoka, K., et al. (2006). Latest blast furnace relining technology at Nippon Steel. *Nippon Steel Technical Report No, 94*, 127–132.

Keii, M., et al. (2010). Structural outline of terrestrial digital broadcasting tower. *Steel Construction Today & Tomorrow, 31*, 5–9.

Kelly, T. D., & Matos, G. R. (2014). Compilers. In *Historical statistics for mineral and material commodities in the United States*. Washington, DC: USGS. <http://minerals.usgs.gov/minerals/pubs/historical-statistics/>.

Kim, S., et al. (2015). Brittle intermetallic compound makes ultrastrong low-density steel with large ductility. *Nature, 518*, 77–79.

King, C. D. (1948). *Seventy-five years of progress in iron and steel.* New York, NY: American Institute of Mining and Metallurgical Engineers.

King, P. (2005). The production and consumption of bar iron in early modern England and Wales. *Economic History Review, 58*, 1–33.

King, P. (2011). The choice of fuel in the eighteenth century iron industry: The Coalbrookdale accounts reconsidered. *Economic History Review, 64*, 132–156.

Knop, K., Hallin, M., & Burström E. (2008). ULCORED SP 12: Concept for minimized CO_2 emission. <http://www.ulcos.org/en/docs/seminars/Ref18%20-%20SP12_Knop_Essen_Final.pdf>.

Kobelco. (2015a). FASTMET process. <http://www.kobelco.co.jp/english/engineering/products/fastmet/index.html>.

Kobelco. (2015b). ITmk3: Process summary. <http://www.kobelco.co.jp/english/engineering/products/ironunit/index.html>.

Kozawa, S., & Tsukihashi, F. (2011). Analysis of global demand for iron source by estimation of in-use steel stock. *ISIJ International, 51*, 320–329.

Kubo, Y., et al. (2012). Steel sheet for the better human life (application for household electrical appliances, OA equipment). *Nippon Steel Technical Reports, 101*, 48–56. <http://www.nssmc.com/en/tech/report/nsc/pdf/NSTR101-09_tech_review-1-3.pdf>.

Kumakura, M. (2013). Advances in steel refining technology and future prospects. *Nippon Steel Technical Report, 104*, 5–12.

Landau, S. B., & Condit, C. W. (1996). *Rise of the New York skyscraper, 1865–1913.* New Haven, CT: Yale University Press.

Laplace Conseil. (2013). Impacts of energy market developments on the steel industry. Paris: Laplace Conseil. <http://www.oecd.org/sti/ind/Item%209.%20Laplace%20-%20Steel%20Energy.pdf>.

Lech Stahlwerke. (2011). Umweltbericht 2009/2010. <http://lech-stahlwerke.de/files/assets/02_news/lsw_ub_2009_2010_ds.pdf>.

Leckie, A. H., Millar, A., & Medley, J. E. (1982). Short- and long-term prospects for energy economy in steelmaking. *Ironmaking and Steelmaking*, *9*, 222–235.

Lee, B., & Sohn, I. (2014). Review of innovative energy savings technology for the electric arc furnace. *JOM*, *66*, 1581–1594.

Lei, Q. (2011). The Development of China's Cement Industry. <http://www.tcma.org.tr/images/file/Cin%20Cimento%20Birligi%20Baskani%20Lei%20QIANZHI%20.pdf>.

Lenzen, M., & Dey, C. (2000). Truncation error in embodied energy analyses of basic iron and steel products. *Energy*, *25*, 577–585.

Lenzen, M., & Treloar, G. (2002). Embodied energy in buildings: Wood versus concrete—reply to Börjesson and Gustavsson. *Energy Policy*, *30*, 249–255.

Li, L., et al. (2014). Comprehensive evaluation of OxyCup process for steelmaking dust treatment based on calculation of mass balance and heat balance. *International Journal of Iron and Steel Research*, *21*, 575–582.

Li, W., & Yang, D. T. (2005). The great leap forward: Anatomy of a central planning disaster. *Journal of Political Economy*, *113*, 840–877.

Lo, F., et al. (1999). *Chinese sustainable development framework summary report.* Tokyo: UNU/IAS.

Lord, W. (1955). *Titanic: A night to remember.* New York, NY: R&W Holt.

Lovis, J. B. (2005). *The blast furnaces of sparrows point: One hundred years of ironmaking on Chesapeake Bay*. Easton, PA: Canal History and Technology.

Luiten, E. E. M. (2001). Beyond energy efficiency. Actors, networks and government intervention in the development of industrial process technologies. Utrecht: Utrecht University.

Lüngen, H. B. (2013). Trends for reducing agents in blast furnace operation. <http://www.dkg.de/akk-vortraege/2013-_-2rd_polnisch_deutsches_symposium/abstract-luengen_reducing-agents.pdf>.

Madias, J. (2014). Electric furnace steelmaking. *Treatise on Process Metallurgy Volume*, *3*, 271–300.

Madureira, N. L. (2012). The iron industry energy transition. *Energy Policy*, *50*, 24–34.

Maillart, R. (1935). Ponts-voûtes en béton armé. De leur développement et de quelques constructions spéciales exécutées en Suisse. *Travaux*, *26*, 64–71.

Mandal, G. K., et al. (2014). A steady state thermal and material balance model for an iron making blast furnace and its validation with operational data. *Transactions of the Indian Institute of Metallurgy*, *67*, 209–221.

Manning, C. P., & Fruehan, R. J. (2001). Emerging technologies for iron and steelmaking. *JOM*, *53*, 2023.

Mao, Z. (1969). *Miscellany of mao zedong thought.* Washington, DC: JPRS.

Markus Engineering Services. (2002). *Cradle-to-gate life cycle inventory: Canadian and US steel production by mill type.* Ottawa: Markus Engineering Services. <http://calculatelca.com/wp-content/themes/athena/images/LCA%20Reports/Steel_Production.pdf>.

Matsui, Y., Terashima, K., & Takahashi, R. (2014). Analysis of scale-up of a shaft furnace by process engineering—From the iron-manufacturing experiment by using Bei-Tetsu in Hippo Tatara. *ISIJ International*, *54*, 1051–1058.

Maupin, M., et al. (2014). *Estimated use of water in the United States in 2010.* Washington, DC: USGS.

Mašlejová, A. (2013). *Utilization of biomass in ironmaking.* Brno: MeTal. 2013. <http://konsys-t.tanger.cz/files/proceedings/12/reports/1744.pdf>.

McCallum, H. D., & McCallum, F. T. (1955). *The wire that fenced the west.* Norman, OK: University of Oklahoma Press.

McCelland, J. (2002). Not all RHFs are created equal. *Direct from MIDREX 2nd Quarter, 2002*, 3–6.

McCloskey, D. N. (1973). *Economic maturity and entrepreneurial decline: British iron and steel 1870–1913*. Cambridge, MA: Harvard University Press.

McKinsey. (2010). *Short-selling the earth?* <www.mckinsey.com/~/.../mckinsey/.../Short_selling_the_Earth_2010.ash>.

McKinsey. (2013a). *Competitiveness and challenges in the steel industry*. <http://www.oecd.org/sti/ind/Item%203.%20McKinsey%20-%20Competitveness%20in%20the%20steel%20industry%20(OECD)%20-%20final.pdf>.

McKinsey. (2013b). *Scarcity and saturation*. <http://www.mckinsey.com/client_service/metals_and_mining/latest_thinking>.

McKinsey. (2013c). *Overcapacities in the steel industry*. <http://www.oecd.org/sti/ind/Item%208.%20%20KcKinsey%20-%20Overcapacities%20in%20the%20steel%20industry%20(OECD)_final.pdf> .

McKinsey. (2014). *Global steel industry perspective—Synthesis version*. <http://www.ats-ffa.org/estad-jsi/downloads/4-Bekaert-a.pdf>.

McKinsey Global Institute. (2013). *Infrastrucure 2013: Global priorities, global insights*. <http://www.ey.com/Publication/vwLUAssets/Infrastructure_2013/$FILE/Infrastructure_2013.pdf>.

McManus, G. J. (1981). Inland's No. 7 start-up more than pushing the right buttons. *Iron Age*, *224*(7) MP-7-MP-16.

McManus, G. J. (1988). Blast furnaces: More heat from the hot end. *Iron Age*, *4*(8), 15–20.

McManus, G. J. (1988). Ironmaking: The next step. *Iron Age*, *4*(8), 29–31.

McManus, G. J. (1989). Coal gets a new shot. *Iron Age*, *5*(1), 31–34.

Metal Suppliers. (2015) Alloy steels 300M. <http://www.suppliersonline.com/propertypages/300M.asp>.

MIDREX, (2012). The economics of longevity. *Direct from MIDREX First Quarter, 2012*, 10–13.

MIDREX. (2014a). MIDREX. <http://www.midrex.com/assets/user/media/MIDREX_Process-Brochure.pdf>.

MIDREX. (2014b). *2013 World direct reduction statistics*. <http://www.midrex.com/assets/user/news/MIDREX_World_DRI_Stats.pdf>.

MIDREX. (2015). The MIDREX process. <http://www.midrex.com/process-technologies/the-midrex-process>.

Millbank, P. (2013). Stainless steel faces subdued global growth. *Insight*, 28–31. May 2013.

Miller, T. W., et al. (1998). Oxygen steelmaking processes. In D. A. Wakelin (Ed.), *The making, shaping and treating of steel, ironmaking volume* (pp. 475–524). Pittsburgh, PA: The AISE Foundation.

Miller, J. R. (1976). The direct reduction of iron ore. *Scientific American*, *235*(1), 68–80.

Minenko, N., et al. (1993). Ural iron before the industrial revolution. In G. Rydén & M. Ågren (Eds.), *Ironmaking in Sweden and Russia: A survey of the social organization of iron production before 1900* (pp. 43–95). Uppsala: Historiska institutionen.

Mining-technology.com. (2014). Ferro giants—The world's biggest iron ore producers. <http://www.mining-technology.com/features/featureferro-giants–the-worlds-biggest-iron-ore-producers-4280601/>.

Misa, T. J. (1995). *A nation of steel: The making of modern America 1865–1925*. Baltimore, MD: The Johns Hopkins University Press.

MMT (Maersk Maritime Technology). (2015). Triple E-vessels. <http://www.maersktechnology.com/stories/stories/pages/triple-evessels.aspx>.

Monteiro, M. A. (2006). Em busca do carvão vegetal barato: o deslocamento de siderúrgicas para a Amazônia. *Novos cadernos NAEA*, *9*(2), 55–97.

Morita, Z., & Emi, T. (Eds.), (2003). *An introduction to iron and steel processing*. Tokyo: Kawasaki Steel 21st Century Foundation. <http://www.jfe-21st-cf.or.jp/index2.html>.

Morris, A. E., Geiger, G., & Fine, H. A. (2011). *Handbook on material and energy balance calculations in materials processing*. Hoboken, NJ: Wiley.

Mourão, J. M. (2011). *NT Minério de Ferro e Pelotas Situação Atual e Tendências 2025*. <http://www.abmbrasil.com.br/epss/arquivos/documentos/2011_4_19_9_7_29_21931.pdf>.

Moynihan, M. C., & Allwood, J. M. (2012). The flow of steel into the construction sector. *Resources, Conservation and Recycling*, *68*, 88–95.

Mushet, D. (1804). Experiments on wootz or Indian steel. *Philosophical Transactions of the Royal Society Series A*, 95:175.

Mussatti, D. C. (1998). *Coke ovens: Industry profile*. Research Triangle Park, NC: USEPA.

Müller, D. B., et al. (2006). Exploring the engine of anthropogenic iron cycles. *Proceedings of the National Academy of Sciences*, *103*, 16111–16116.

Naito, M., Takeda, K., & Matsui, Y. (2015). Ironmaking technology for the last 100 years. *ISIJ International*, *55*(1), 7–35.

Nasaw, D. (2006). *Andrew Carnegie*. New York, NY: Penguin Books.

NBS. (2013). *China statistical yearbook*. Beijing: NBS.

NBS (National Bureau of Statisics). (2000). *China statistical yearbook*. Beijing: NBS.

Needham, J. (1964). *The development of iron and steel in China*. London: The Newcomen Society.

Newby, F. (Ed.), (2001). *Early reinforced concrete*. Burlington, VT: Ashgate Publishing.

Nogami, H., et al. (2006). Analysis on material and energy balances of ironmaking systems on blast furnace operations with metallic charging, top gas recycling and natural gas injection. *ISIJ International*, *46*, 1759–1766.

Nomura, S., & Callcott, T. G. (2011). Maximum rates of pulverized coal injection in ironmaking blast furnaces. *ISIJ International*, *51*, 1033–1043.

Norgate, T., & Landberg, D. (2009). Environmental and economic aspects of charcoal use in steelmaking. *ISIJ International*, *49*, 587–595.

NSA (National Slag Association). (2015). Blast furnace slag. <http://www.nationalslag.org/blast-furnace-slag>.

NSSE (Nippon Steel & Sumikin Engineering). (2013). The track record for 50 years. <http://www.nsengi.net/pdf/02_mmmm/news.pdf>.

Nucor. (2014). Nucor today. <www.nucor.com/media/IR-March2014InvestorPresentation.pptx>.

Nyboer, J., & Bennett, M. (2014). *Energy use and related data: Canadian iron and steel and ferroalloy manufacturing industries 1990 to 2012*. Burnaby, BC: Canadian Industrial Energy End-use Data and Analysis Centre.

Oda, J., et al. (2012). International comparisons of energy efficiency in power, steel, and cement industries. *Energy Policy*, *44*, 118–129.

OECD. (2001). *An initial view on methodologies for emissions baselines: Iron and steel case study*. Paris: OECD.

O'Hara, M. (2014). An insight into U.S. Steel's transformation. <http://marketrealist.com/2014/11/insight-u-s-steels-transformation/>.

Ogaki, Y., et al. (2001). Recycling of waste plastic packaging in a blast furnace system. <http://www.jfe-steel.co.jp/archives/en/nkk_giho/84/pdf/84_01.pdf>.

Ogawa, S., et al. (2012). Progress and prospect of rolling technology. *Nippon Steel Technical Report*, *101*, 95–104. <http://www.nssmc.com/en/tech/report/nsc/pdf/NSTR101-14_tech_review-2-3.pdf>.

Ogawa, T., Sellan, R., & Ruscio, E. (2011). Jumbo size 420t twin DC FastArc® EAF at Tokyo Steel. *Millennium Steel*, *2011*, 52–58.

Ogura, S. et al. (2014). Environmental conservation and energy saving activities in JFE Steel. JFE Technical Report No. 19 March 2014. <http://www.jfe-steel.co.jp/en/research/report/019/pdf/019-19.pdf>.

Ohashi, N. (1992). Modern steelmaking. *American Scientist*, *80*, 540–555.

Okumura, H. (1994). Recent trends and future prospects of continuous casting technology. *Nippon Steel Technical Report, 61*, 9–14.

Okuno, Y. (2006). Prospects of iron and steel production and progress of blast furnace route in China. *Nippon Steel Technical Reports, 94*, 16–22. <http://www.nssmc.com/en/tech/report/nsc/pdf/n9403.pdf>.

Olsson., F. (2007). *Järnhanteringens dynamic: Produktion, lokalisering och agglomerationer i Bergslagen och Mellansverige 1368–1910*. Umeå: Umeå Studies in Economic History.

Osborn, F. (1952). *The story of the mushets*. London: Thomas Nelson & Sons.

Osborne, D. (Ed.), (2013). *The coal handbook: Towards cleaner production*. Cambridge: Woodhead Publishing.

Outotec. (2015a). Outotec© sintering technologies. <http://www.outotec.com/en/About-us/>.

Outotec. (2015b). Outotec© pelletizing technologies. <http://www.outotec.com/en/About-us/>.

Pacey, A. (1992). *Technology in world civilization*. Cambridge, MA: MIT Press.

Palgrave Macmillan. (Eds.), (2013). *International Historical Statistics*. <http://www.palgrave-connect.com/pc/connect/archives/ihs.html>.

Palmer, T. (2014). Climate forecasting: Build high-resolution global climate models. *Nature, 515*, 338–339.

Pardo, N., Moya, J. A., & Vatopoulos, K. (2012). *Perspective scenarios on energy efficiency and CO_2 emissions in the EU iron & steel industry*. Brussels: European Commission. <https://ec.europa.eu/jrc/sites/default/files/ldna25543enn.pdf>.

Park, J., & Rehren, T. (2011). Large-scale 2nd and 3rd century AD bloomery iron smelting in Korea. *Journal of Archaeological Science, 38*, 1180–1190.

Paskoff, P. F. (Ed.), (1989). *Iron and steel in the nineteenth century*. New York, NY: Facts on File.

Pauliuk, S., et al. (2013). The steel scrap age. *Environmental Science & Technology, 47*, 3348–3354.

Pauliuk, S., Wang, T., & Müller, D. B. (2013). Steel all over the world: Estimating in-use stocks of iron for 200 countries. *Resources, Conservation and Recycling, 71*, 22–30.

Paynter, S. (2006). Regional variations in bloomer smelting slag of the Iron Age and Romano-British periods. *Archaeometry, 48*, 271–292.

PE International. (2012). *Comparative life cycle assessment of aluminum and steel truck wheels*. Boston, MA: PE International. <http://www.alcoawheels.com/alcoawheels/north_america/en/pdf/Alcoa_Comparative_LCA_of_Truck_Wheels_with_CR_statement.pdf>.

Peacey, J. G., & Davenport, W. G. (1979). *The iron blast furnace*. Oxford: Pergamon Press.

Peláez-Samaniegoa, M. R. (2008). Improvements of Brazilian carbonization industry as part of the creation of a global biomass economy. *Renewable and Sustainable Energy Reviews, 12*, 1063–1086.

Pereira, B. L. C., et al. (2012). Quality of wood and charcoal from *Eucalyptus* clones for ironmaster use. *International Journal of Forestry Research*. http://dx.doi.org/10.1155/2012/523025.

Perttula, J. (2004). Wootz Damascus steel of ancient orient. *Scandinavian Journal of Metallurgy, 33*, 92–97.

Pfeifer, H., & Kirschen M. (2002). Thermodynamic analysis of EAF energy and comparison with a statistical model of electric energy demand. <http://citeseerx.ist.psu.edu/viewdoc/download?doi=10.1.1.195.4061&rep=rep1&type=pdf>.

Pfeifer, H. C., Sousa, L. G. & Silva T. T. (2012). Design of the charcoal blast furnace – Differences to the Coke BF. Paper presented at the 6th International Congress of the Science and Technology in Iron Making, Rio de Janeiro, October 14-18, 2012. <http://www.abmbrasil.com.br/cim/download/Henrique%20C%20Pleeifer%20Design%20of%20Charcoal%20Blast.pdf>.

Polinares. (2012). *Fact sheet: Iron ore*. Brussels: European Commission.

Polmar, N., & Allen, T. B. (2012). *World War II: The encyclopedia of the war years, 1941–1945*. New York, NY: Dover.

Porter, H. C. (1924). *Coal carbonization*. New York, NY: The Chemical Catalog Company.

POSCO. (2013). POSCO renovates world's biggest furnace at Gwangyang Steelworks. *POSCO Official Blog* June 13, 2013. <http://globalblog.posco.com/posco-renovates-worlds-biggest-furnace-at-gwangyang-steelworks/>.

Poveromo, J. J. (2006). Agglomeration processes—pelletizing and sintering. In *Industrial minerals & rocks—Commodities, markets and uses* (pp. 1391–1404). Littleton: Society for Mining, Metallurgy and Exploration (pp. 1391–1404).

Prime Research. (2014). *World Car Trends 2014 "Smart Efficiency and Digital Intelligence"*. <http://www.wcoty.com/files/Library/2/World%20Car%20of%20the%20Year%20Study_Global%20Automotive%20Trends%202014.pdf>.

Protásio, T. et al. (2014). Qualidade e avaliação energética do carvão vegetal dos resíduos do coco babaçu para uso siderúrgico. *Ciência e Agrotecnologia, 38*, 435–444.

Ransom, P. J. G. (1989). *The Victorian railway and how it evolved*. London: William Heinemann.

Recycling International. (2014). Plunge in US steel scrap exports to Turkey and China. <http://www.recyclinginternational.com/recycling-news/8388/ferrous-metals/turkey/plunge-us-steel-scrap-exports-turkey-and-china>.

Rehren, T., et al. (2013). 5,000 years old Egyptian iron beads made from hammered meteoritic iron. *Journal of Archaeological Science, 40*, 4785–4792.

Remus, R., et al. (2013). Best available techniques (BAT) reference document for iron and steel production. <http://eippcb.jrc.ec.europa.eu/reference/BREF/IS_Adopted_03_2012.pdf>.

Riedel, G. (1994). *Der siemens-martin-ofen: Rückblick auf eine stahlepoche*. Düsseldorf: Stahleisen Verlag.

Rio Tinto. (2014). HIsmelt. <http://www.riotinto.com/ironore/hismelt-4724.aspx>.

Rousmaniere, P., & Raj, N. (2007). Shipbreaking in the developing world: Problems and prospects. *International Journal of Occupational and Environmental Health, 13*, 359–368.

Rousset, P., et al. (2011). Pressure effect on the quality of eucalyptus wood charcoal for the steel industry: A statistical analysis approach. *Fuel Processing Technology, 92*, 1890–1897.

Ryan, M. P., et al. (2002). Why stainless steel corrodes. *Nature, 415*, 770–774.

Rydén, G., & Ågren, M. (Eds.), (1993). *Ironmaking in Sweden and Russia: A survey of the social organization of iron production before 1900*. Uppsala: Historiska institutionen.

Ryman, C., et al. (2004). Modelling of the blast furnace process with a view to optimize the steel plant energy system. In: Proceedings: 2nd International Conference & Exhibition on New Developments in Metallurgical Process Technology, Riva del Garda, Italy, September 19–21, 2004. Milano: Associazione Italiana di Metallurgia.

Salzgitter Mannesmannröhren-Werke. (2015). Stahlrohre von Mannesmann—Innovation und Tradition. <http://www.mrw.de/>.

Samajdar, C. (2012). Reduction in specific energy consumption in steel industry—with special reference to Indian steel industry. *Energy and Environmental Engineering Journal, 1*, 104–107.

Sampaio, R. S. (2005). Large-scale charcoal production to reduce CO_2 emissions and improve quality in the coal based ironmaking industry. Paper presented the *Workshop and Business Forum on Sustainable Biomass Production for the World Market*, Campinas, 30.11-3.12 2005. <http://www.bioenergytrade.org/downloads/sampaionovdec05.pdf>.

Sasada, T., & Chunag, A. (2014). Irom smelting in the nomadic empire of Xiongnu in ancient Mongolia. *ISIJ International, 54*, 1017–1023.

Schmidt, C. M., & Döhrn R. Stahl als unverzichtbarer Eckpfeiler der deutschen Industrie. <http://www.stahl-online.de/wp-content/uploads/2013/10/20140414_Stahl_unter_Strom_WEB.pdf>.

Schmidt, P., & Avery, D. H. (1978). Complex iron smelting and prehistoric culture in Tanzania. *Science, 201*, 1085–1089.

Schmidt, P. R., & Childs, S. T. (1995). Ancient African iron production. *American Scientist, 83*, 524–533.

Schmöle, P., Lüngen, H. B., & Noldin J. H. (2014). Trends in iron-making given the new reality of iron ore and coal resources. <http://www.abmbrasil.com.br/cim/download/Jos%C3%A9%20Henrique%20Noldin%20J%C3%BAnior%20-%20Trends%20in%20iron-making%20given%20the%20new%20reality%20of%20ore%20and%20coal%20resources.pdf>.

Schnatterly, J. (2008). Watching our weight: Steel content of North American auto. <http://www.autosteel.org/~/media/Files/Autosteel/Great%20Designs%20in%20Steel/GDIS%202010/10%20-%20Watching%20Our%20Weight%20%20Steel%20Content%20of%20N%20American%20Auto.pdf>.

Schneider, W. (2000). *Continuous casting*. New York, NY: John Wiley.

Schrewe, H. F. (1991). *Continuous casting of steel: fundamental principles and practice*. Düsseldorf: Stahl und Eisen.

Schumpeter, J. A. (1942). *Capitalism, socialism, and democracy*. New York, NY: Harper.

Scientific American. (1913). The electric production of steel. *Scientific American*, 88–89. August 2, 1913.

Scott, D. A. (2002). *Copper and bronze in art: Corrosion, colorants, conservation*. Los Angeles: Getty Conservation Institute.

SCS Global Services. (2013). *Environmental life cycle assessment of southern yellow pine wood and North American galvanized steel utility distribution poles*. Emeryville, CA: SCS Global Services. <http://www.steel.org/~/media/Files/SMDI/Construction/Utility%20Poles/Utility%20Pole%20-%20Steel%20vs%20Wood%20LCA%20Study%20Executive%20Summary.pdf>

Seitz, F. (2014). *Gustave Eiffel. Le triomphe de l'ingénieur*. Paris: Armand Colin.

Sexton, A. H. (1897). *Fuel and refractory materials*. London: Vlackie and Son.

Shaeffer, R. E. (1992). *Reinforced concrete: Preliminary design for architects and builders*. New York, NY: McGraw-Hill.

Shepard, R. R. (2004). Steel's workhorse: The basic oxygen furnace. *Inspection Trends*, 7(1), 21–23. <https://app.aws.org/itrends/2004/01/it0104-21.pdf>.

Shinotake, A., et al. (2004). Blast furnace campaign life relating to the productivity. *La Revue de Métallurgie, 2004*, 203–209.

Ship Cruise. (2015). Cruise ship cost to build. <http://www.shipcruise.org/how-much-does-a-cruise-ship-cost/>.

SIA. (2014). Semiconductor industry posts record sales in 2013. <http://www.semiconductors.org/news/2014/02/03/global_sales_report_2013/semiconductor_industry_posts_record_sales_in_2013/>.

Siemens VAI. (2006). SIMETALCIS BF—Solutions for blast furnaces. <https://www.industry.siemens.com/datapool/industry/industrysolutions/metals/simetal/en/SIMETAL-BF-en.pdf>.

Siemens VAI. (2008). SIMETAL^{CIS} hot blast stoves. <https://www.industry.siemens.com/datapool/industry/industrysolutions/metals/simetal/en/SIMETAL-Hot-Blast-Stoves.pdf>.

Siemens VAI. (2011). SOMETAL corex technology. <https://www.industry.siemens.com/datapool/industry/industrysolutions/metals/simetal/en/SIMETAL-Corex-technology-en.pdf>.

Siemens VAI. (2012). *High productive steelmaking with SIMETAL EAF ultimate*. <http://www.industry.siemens.com/datapool/industry/industrysolutions/metals/simetal/en/High-productive-steelmaking-with-SIMETAL-EAF-Ultimate-COLAKOGLU-en.pdf>.

SkyscraperPage.com (2015). World's tallest buildings 2015. <http://skyscraperpage.com/diagrams/?searchID=200>.

Smil, V. (1994). *Energy in world history*. Boulder, CO: Westview Press.

Smil, V. (2000). China's great famine: 40 years later. *British Medical Journal, 7225*, 1619–1621.
Smil, V. (2004). *China's past, China's future*. London: RoutledgeCurzon.
Smil, V. (2005). *Creating the twentieth century: Technical innovations of 1867–1914 and their lasting impact*. New York, NY: Oxford University Press.
Smil, V. (2006). *Transforming the 20th century: Technical innovations and their consequences*. New York, NY: Oxford University Press.
Smil, V. (2010). *Prime movers of globalization*. Cambridge, MA: MIT Press.
Smil, V. (2013). *Making the modern world*. Chichester: John Wiley.
Smil, V. (2015). *Power density*. Cambridge, MA: MIT Press.
Smil, V., Nachman, P., & Long, T.V. (1983). *Energy analysis and agriculture: An application to U.S. corn production*. Boulder, CO: Westview Press.
Smithson, D. J., & Sheridan, A. T. (1975). Energy use in mill areas. *Ironmaking and Steelmaking, 4*, 286–294.
Sohn, I., & Fruehan, R. J. (2006). The reduction of iron oxides by volatiles in a rotary hearth furnace process. *Metallurgical and Materials Transactions B, 37*, 223–229.
SRI. (2015). Steel is North America's #1 recycled material. <http://www.recycle-steel.org/>.
SRI (Steel Recycling Institute). (2014). 2013 steel recycling rates. <http://www.recycle-steel.org/~/media/Files/SRI/Releases/Steel%20Recycling%20Rates%20Sheet.pdf?la=en>.
Stahlinstitut der VDEh. (2013). *Beitrag der Stahlindustrie zu Nachhaltigkeit, Ressourcen- und Energieeffizienz*. <http://www.stahl-online.de/wp-content/uploads/2013/10/Beitrag-der-Stahlindustrie-zu-Nachhaltigkeit-Ressourcen-und-Energieeffizienz.pdf>.
Stahlinstitut VDEh. (2014). *Fakten zur Stahlindustrie in Deutschland*. <http://www.stahl-online.de/wp-content/uploads/2013/12/Fakten_Stahlindustrie_2014-04_6.pdf>.
Starratt, F. W. (1960). LD … in the beginning. *Journal of Metals, 12*, 528–530.
Steel Benchmarker. (2015). Price history: Tables and charts. <http://steelbenchmarker.com/files/history.pdf>.
Steel Framing Alliance. (2007). *A builder's guide to steel frame construction*. <http://www.steelframing.org/PDF/SFA_Framing_Guide_final%202.pdf>.
Steel Times International. (2013). HIsmelt plant goes to China. <http://www.steeltimesint.com/news/view/hismelt-plant-goes-to-china>.
Steiger, R. W. (1999). Edison's concrete dream. <http://www.concreteconstruction.net/Images/Edison's%20Concrete%20Dream_tcm45-343558.pdf>.
Straker, E. (1969). *Wealden iron*. New York, NY: Augustus M. Kelley.
Stubbles, J. (2000). *Energy use in the U.S. steel industry: An historical perspective and future opportunities*. Columbia, MD: Energetics.
Stubbles, J. (2000). EAF steelmaking—past, present and future. *Direct from MIDREX, 3*, 3–4.
Stubbles, J. (2015). The Basic Oxygen Steelmaking (BOS) Process. <http://www.steel.org/Making%20Steel/How%20Its%20Made/Processes/Processes%20Info/The%20Basic%20Oxygen%20Steelmaking%20Process.aspx>.
Sugawara, T., et al. (1986). Construction and operation of No. 5 blast furnace, Fukuyama Works, Nippon Kokan KK. *Ironmaking and Steelmaking, 3*, 241–251.
Sullivan, D.E. (2005). Metal stocks in use in the United States. <http://pubs.usgs.gov/fs/2005/3090/2005-3090.pdf>.
Sundholm, J. L., et al. (1999). Manufacture of metallurgical coke and recovery of coal chemicals. In D. A. Wakelin (Ed.), *The making, shaping and treating of steel, ironmaking volume* (pp. 381–545). Pittsburgh, PA: The AISE Foundation.
Suopajärvi, H., & Fabritius, T. (2013). Towards more sustainable ironmaking—An analysis of energy wood availability in Finland and the economics of charcoal production. *Sustainability, 2013*, 1188–1207.

Svensson, E., et al. (2009). The crofter and the iron works: The material culture of structural crisis, identity and making a living on the edge. *International Journal of Historical Archaeology, 13*, 183–205.

Szekely, J. (1987). Can advanced technology save the U.S. steel industry? *Scientific American, 257*(1), 34–41.

Takahashi, M., et al. (2012). Steels and their applications for life satisfaction and transportation. *Nippon Steel Technical Report, 101*, 27–36.

Takahashi, M. (2015). Sheet steel technology for the last 100 years: Progress in sheet steels in hand with the automotive industry. *ISIJ International, 55*, 79–88.

Takahashi, M., Hongu, A., & M. Honda (1994). Recent advances in electric arc furnaces for steelmaking. *Nippon Steel Technical Report* 61:58–64.

Takamatsu, N., et al. (2012). Development of iron-making technology. *Nippon Steel Technical Report, 101*, 79–88.

Takamatsu, N., et al. (2014). Steel recycling circuit in the world. *Tetsu to hagane, 100*, 740–749.

Takeuchi, H., et al. (1994). Production of stainless steel strip by twin-drum strip casting process. *Nippon Steel Technical Report, 61*, 46–51.

Tang, R. (2010). *China's steel industry and its impact on the United States: Issues for congress.* <http://digitalcommons.ilr.cornell.edu/cgi/viewcontent.cgi?article=1761&context=key_workplace>.

Tanner, A. H. (1998). *Continuous casting: A revolution in steel.* Fort Lauderdale: Write Stuff Enterprises.

Tassava, C. (2008). The American Economy during World War II. *EH.Net Encyclopedia*, Whaples R. (Ed.) February 10, 2008. <http://eh.net/encyclopedia/article/tassava.WWII>.

Taylor, F. W. (1907). *On the art of cutting metals.* New York, NY: ASME.

Taylor, F. W. (1911). *The principles of scientific management.* New York, NY: Harper and Brothers.

Team Stainless. (2014). Stainless steel. <http://www.teamstainless.org/>.

Temin, P. (1964). *Iron and steel in nineteenth century america.* Cambridge, MA: MIT Press.

Tezuka, H. (2014). Voluntary actions in the Japanese steel industry. <http://www.meti.go.jp/policy/energy_environment/kankyou_keizai/va/seika/140806/5_2.pdf>.

Thakkar, V., et al. (2008). Life cycle assessment of some indian steel industries with special reference to the climate change. *Journal of Environmental Research and Development, 2*, 773–782.

Thomas, J. (Ed.), (1979). *Energy analysis.* Boulder, CO: Westview Press.

Thomsen, C. J. (1836). *Ledetraad til nordisk oldkyndighed.* Copenhagen: L. Mellers.

ThyssenKrupp. (2003). 30 Jahre Hochofen in Duisburg-Schwelgern—Vom schwarzen Riesen zum Hightech-Giganten. <http://www.thyssenkrupp.com/de/presse/art_detail.html&eid=tk_pnid751>.

ThyssenKrupp. (2008). Fifth furnace campaign can begin: Blast furnace 1 in Duisburg-Schwelgern to restart operation in April after modernization. <http://www.thyssenkrupp.com/en/presse/art_detail.html&device=printer&eid=TKBase_1205843849010_520600257>.

ThyssenKrupp. (2014). Größter Hochofen Europas angeblasen: "Schwelgern 2" erschmilzt wieder Roheisen. <https://www.thyssenkrupp-steel-europe.com/de/presse/pressemitteilungen/pressemitteilung-6610.html>.

ThyssenKrupp. (2015). ThyssenKrupp AG. <https://www.thyssenkrupp.com/>.

Toulouevski, Y. N., & Zinurov, I. Z. (2010). *Innovation in electric arc furnaces.* Berlin: Springer.

Tovarovskiy, I. G. (2013). Substitution of coke and energy savings in blast furnaces. *Energy Science and Technology, 6*, 4–13.

Tunc, M., Camdali, U., & Arasil, G. (2012). Mass analysis of an electric furnace at a steel company in Turkey. *Metallurgis, 56*, 253–261.

Turak, T. (1986). *William Le Baron Jenney: A pioneer of modern architecture.* Ann Arbor, MI: UMI Research Press.

Tylecote, R. F., Austin, J. N., & Wraith, A. E. (1971). The mechanism of the bloomery process in shaft furnaces. *Journal of the Iron and Steel Institute, 209*, 342–363.

Uemori, R., et al. (2012). Steels for energy production and transport. *Nippon Steel Technical Report, 101*, 68–78.

Uemori, R., et al. (2012). Steels for marine transportation and construction. *Nippon Steel Technical Report, 101*, 37–47.

Uhlig, A. (2011). Charcoal production in Brazil: Does it pass the sustainability bar? Paper presented at *Charcoal Symposium*, Arusha, Tanzania, June 15, 2011. <http://www.charcoalproject.org/wp-content/uploads/2011/08/10_Uhlig_Brazil_stats.pdf>.

Uhlmann, J., & Heinrich, P. (1987). *The soul of fire: How charcoal changed the world*. Pompano Beach, FL: University Books.

UNEP (United Nations Environment Programme). (2011). *Recycling rates of metals: A status report* (<http://www.unep.org/resourcepanel/Portals/24102/PDFs/Metals_Recycling_Rates_110412-1.pdf>). Nairobi: UNEP. <http://www.unep.org/resourcepanel/Portals/24102/PDFs/Metals_Recycling_Rates_110412-1.pdf>

USBC (US Bureau of the Census). (1975). *Historical statistics of the United States: Colonial times to 1970*. Washington, DC: US Department of Commerce.

USDOE (US Department of Energy). (2013). *Lightweight Materials: 2012 Annual Progress Report*. <https://www1.eere.energy.gov/vehiclesandfuels/pdfs/program/2012_lightweight_materials.pdf>.

USDOT (US Department of Transportation). (2015). Motor vehicles scrapped: Table 4-58. <http://www.rita.dot.gov/bts/sites/rita.dot.gov.bts/files/publications/national_transportation_statistics/index.html>.

USEIA. (2015). How much electricity does an American home use? <http://www.eia.gov/tools/faqs/faq.cfm?id=97&t=3>.

USEPA. (2007). Energy trends in selected manufacturing sectors: Opportunities and challenges for environmentally preferable outcomes. <http://www.epa.gov/sectors/pdf/energy/report.pdf>.

USEPA. (2008). *2008 Sector performance report*. Washington, DC: USEPA. <http://www.epa.gov/sectors/pdf/2008/2008-sector-report-508-full.pdf>.

USEPA. (2012). *Available and emerging technologies for reducing greenhouse gas emissions from the iron and steel industry*. Research Triangle Park, NC: USEPA. <http://www.epa.gov/nsr/ghgdocs/ironsteel.pdf>.

USEPA. (2014). Municipal solid waste generation, recycling, and disposal in the United States: Facts and figures for 2012. <http://www.epa.gov/osw/nonhaz/municipal/pubs/2012_msw_fs.pdf>.

USGS. (2012). Building safer structures. <http://earthquake.usgs.gov/learn/publications/saferstructures/>.

USGS (US Geological Survey). (2014). Mineral commodity summaries 2014. <http://minerals.usgs.gov/minerals/pubs/mcs/2014/mcs2014.pdf>.

USNRC (US Nuclear Regulatory Council). (2014). Fact sheet on reactor pressure vessel issues. <http://www.nrc.gov/reading-rm/doc-collections/fact-sheets/prv.html>.

USS (US Steel). (2015). Gary works. <https://www.ussteel.com/uss/portal/home/aboutus/facilities/company-facilities-garyworks-garyindi/>.

Vadenbo, C. O., Boesch, M. E., & Hellweg, S. (2013). Life cycle assessment model for the use of alternative resources in Ironmaking. *Journal of Industrial Ecology, 17*, 363–374.

Vale. (2015). Valemax. <http://www.vale.com/EN/initiatives/innovation/valemax/pages/default.aspx>.

Valia, H. S. (2014). *Coke production for blast furnace ironmaking*. Washington, DC: AISI. <https://www.steel.org/Making%20Steel/How%20Its%20Made/Processes/Processes%20Info/Coke%20Production%20For%20Blast%20Furnace%20Ironmaking.aspx>.

Van Noten, F., & Raymaekers, J. (1988). Early iron smelting in Central Africa. *Scientific American, 258*, 104–111.

VDEh. (2013). *Blast furnaces worldwide. VDEh PLANTFACTS.* <http://en.stahl-online.de/wp-content/uploads/2013/09/PLANTSFACTSE.pdf>.

Verbraeck, A. (Ed.), (1976). *The energy accounting of materials, products, processes and services.* Rotterdam: TNO (Netherlands Institute for Applied Scientific Research).

Verhoeven, J. D. (1987). Damascus steel. Part I: Indian wootz steel. *Metallography, 20,* 145–151.

Verhoeven, J. D. (2001). The mystery of Damascus blades. *Scientific American, 284*(1), 74–79.

Verhoeven, J. D., Pendray, A. H., & Dauksch, W. E. (1998). The key role of impurities in ancient Damascus steel blades. *Journal of Metals,* 58–62. September 1998.

Voysey, H. W. (1832). Description of the native manufacture of steel in southern India. *Journal of the Asiatic Society of Bengal* 1L245-247.

Wagner, D. B. (1993). *Iron and steel in ancient China.* Leiden: E.J. Brill.

Wagner, D. B. (2013). *The traditional Chinese iron industry and its modern fate.* London: Routledge.

Wakelin, D. H., & Fruehan, R. J. (Eds.), (1999). *Making, shaping and treating of steel (Iron Making).* Pittsburgh, PA: The AISE Steel Foundation.

Walker, R. D. (1985). *Modern ironmaking methods.* Brookfield, VT: Gower Publishing.

Walsh, S. (2011). Iron ore. Sydney: Rio Tinto Investor Seminar, November 28, 2011. <http://www.riotinto.com/documents/111128_RIo_Tinto_Investor_Seminar_slides.pdf>.

Wang, C., et al. (2008). A model on CO_2 emission reduction in integrated steelmaking by optimization methods. *International Journal of Energy Research, 32,* 1092–1106.

Wang, T., et al. (2007). Forging the anthropogenic iron cycle. *Environmental Science & Technology, 41,* 5120–5129.

Warren, K. (2008). *Bethlehem steel: Builder and arsenal of America.* Pittsburgh, PA: University of Pittsburgh Press.

Washlaski, R. A. (2008). Manufacture of coke at Salem No. 1 mine coke works. <http://patheoldminer.rootsweb.ancestry.com/coke2.html>.

Wayman, M. L. (1988). The early use of iron in Arctic North America. *JOM, 40,* 44–45.

WAZ. (2013). Duisburgs "Schwarzer Riese"—Hochofen Schwelgern 1 produziert seit 40 Jahren <http://www.derwesten.de/staedte/duisburg/duisburgs-schwarzer-riese-hochofen-schwelgern-1-produziert-seit-40-jahren-id7609277.html>.

WCA (World Coal Association). (2014). Coal statistics. <http://www.worldcoal.org/resources/coal-statistics/>.

WCA (World Coal Association), 2015. Coal Statistics. <http://www.worldcoal.org/resources/coal-statistics/>.

Wengenroth, U. (1994). *Enterprise and technology: The German and British steel industries, 1865–1895.* Cambridge: Cambridge University Press.

White Star Line. (2008). *Titanic* and other White Star Line ships. <http://www.titanic-whitestarships.com/>.

Wight, J. K., & MacGregor, J. G. (2011). *Reinforced concrete: Mechanics and design.* Englewood Cliffs, NJ: Prentice Hall.

Williams, A. (2009). A note on liquid iron in medieval Europe. *Ambix, 56,* 68–75.

Williams, M. (2006). *Deforesting the earth: From prehistory to global crisis.* Chicago: Chicago University Press.

Wirtschaftsvereinigung Stahl. (2013). *Energiewende beginnt mit Stahl.* <http://www.stahl-online.de/wp-content/uploads/2013/10/121205_Energiewende_beginnt_mit_Stahl.pdf>.

WISDRI. (2012). New high production indexes achieved in 5,800 m^3 blast furnace of Shagang. <http://www.wisdri.com/en/news/detail.aspx?cid=67&id=12608>.

Wood, P. (2015). Sign of the times: Sparrows Point blast furnace demolished. *The Baltimore Sun,* January 28, 2015. <http://www.baltimoresun.com/news/maryland/baltimore-county/dundalk/bs-md-co-sparrows-point-implosion-20150128-story.html>.

WorldAutoSteel. (2011). *Future steel vehicle*. <http://www.worldautosteel.org/projects/future-steel-vehicle/phase-2-results/>.

Worrell, E., et al. (2008). *World best practice energy intensity values for selected industrial sectors*. Berkeley, CA: Ernest Orlando Lawrence Berkeley National Laboratory. <https://ies.lbl.gov/sites/all/files/industrial_best_practice_en_1.pdf>.

Worrell, E., et al. (2010). *Energy efficiency improvement and cost saving opportunities for the U.S. iron and steel industry an ENERGY STAR® guide for energy*. Berkeley, CA: Ernest Orlando Lawrence Berkeley National Laboratory.

WSA. (1982). Steel statistical yearbook. <http://www.worldsteel.org/statistics/statistics-archive/yearbook-archive.html>.

WSA. (1990). Steel statistical yearbook. <http://www.worldsteel.org/statistics/statistics-archive/yearbook-archive.html>.

WSA. (2001). Steel statistical yearbook. <http://www.worldsteel.org/statistics/statistics-archive/yearbook-archive.html>.

WSA. (2009). The three Rs of sustainable steel. <https://www.steel.org/~/media/Files/SMDI/Sustainability/3rs.pdf?la=en>.

WSA. (2011a). Steel and raw materials. <http://www.steelforpackaging.org/uploads/ModuleXtender/Themesslides/13/Fact_sheet_Raw_materials2011.pdf>.

WSA. (2011b). Steel food cans. <http://www.worldsteel.org/dms/internetDocumentList/case-studies/Food-cans-case-study2011/document/Food%20cans%20case%20study2011.pdf>.

WSA. (2011c). *Life cycle assessment methodology report*. Brussels: WSA. <http://www.worldsteel.org/dms/internetDocumentList/bookshop/LCA-Methodology-Report/document/LCA%20Methodology%20Report.pdf>.

WSA. (2012a). *Steel solutions in the green economy: Wind turbines*. Brussels: WSA. <https://www.worldsteel.org/dms/internetDocumentList/bookshop/worldsteel-wind-turbines-web/document/Steel%20solutions%20in%20the%20green%20economy:%20Wind%20turbines.pdf>.

WSA. (2012b). *Sustainable steel: At the core of a green economy*. Brussels: WSA. <http://www.worldsteel.org/dms/internetDocumentList/bookshop/Sustainable-steel-at-the-core-of-a-green-economy/document/Sustainable-steel-at-the-core-of-a-green-economy.pdf>.

WSA. (2014a). *Steel statistical yearbook 2014*. <http://www.worldsteel.org/dms/internet-DocumentList/statistics-archive/yearbook-archive/Steel-Statistical-Yearbook-2014/document/Steel-Statistical-Yearbook-2014.pdf>.

WSA. (2014b). Steel industry by-products. <http://www.worldsteel.org/publications/fact-sheets/content/01/text_files/file/document/Fact_By-products_2014.pdf>.

WSA. (2015). January 2015 crude steel production. <http://www.worldsteel.org/media-centre/press-releases/2015/January-2015-crude-steel-production-for-the-65-countries-reporting-to-worldsteel.html>.

WSC (World Shipping Council). (2015). Containers. <http://www.worldshipping.org/about-the-industry/containers>.

Wu, Y., & Ling, H. C. (1963). *Economic development and the use of energy resources in communist China*. New York, NY: Praeger.

Yamaguchi, J., Nakashima, T., & Sawai, T. (2013). Change and development of continuous casting technology. *Nippon Steel Technical Report*, *104*, 13–20.

Yamazaki, Y. (2012). Gasification reactions of metallurgical coke and its application—Improvement of carbon use efficiency in blast furnace. <http://www.intechopen.com/books/gasification-for-practical-applications/gasification-reactions-of-metallurgical-coke-and-its-application-improvement-of-carbon-use-efficiency>.

Yasuba, Y. (1996). Did Japan ever suffer from a shortage of natural resources before World War II? *The Journal of Economic History*, *56*, 543–560.

Yellishetty, M., & Mudd, G. M. (2014). Substance flow analysis of steel and long term sustainability of iron ore resources in Australia, Brazil, China and India. *Journal of Cleaner Production*, *84*, 400–410.

Yetisken, Y., Camdali, U., & Ekmekci, I. (2013). Cost and energy analysis for optimization of charging materials for steelmaking in EAF and LF as a system. *Metallurgist*, *57*, 378–388.

Yin, X., & Chen, W. (2013). Trends and development of steel demand in China: A bottom-up analysis. *Resources Policy*, *38*, 407–415.

Yonekura, S. (1994). *The Japanese iron and steel industry, 1850–1990: Continuity and discontinuity*. New York, NY: St. Martin's Press.

Zambas, K. (1992). Structural repairs to the monuments of the Acropolis—The Parthenon. *Civil Engineering*, *92*, 166–176.

Zangato, E., & Holl, A. F. C. (2010). On the iron front: New evidence from North-central Africa. *Journal of African Archaeology*, *8*, 7–23.

Zhang, J., & Wang, G. (2009). Energy saving technologies and productive efficiency in the Chinese iron and steel sector. *Energy*, *33*, 525–537.

Zhang, S., & Yu, Z. (2009). Practice and concept for extending blast furnace campaign life at WISCO. *Journal of Iron and Steel International*, *16*, 830–835.

Zschokke, B. (1924). Du damassé et des lames de Damas. *Revue de Metallurgie*, *21*, 639–669.

INDEX

Note: Page numbers followed by "*f*" and "*t*" refer to figures and tables, respectively.

E

F

G

H

I